Marine Ecology: Concepts and Applications

Marine Ecology: Concepts and Applications

Ashley Waters

SYRAWOOD
PUBLISHING HOUSE
New York

Published by Syrawood Publishing House,
750 Third Avenue, 9th Floor,
New York, NY 10017, USA
www.syrawoodpublishinghouse.com

Marine Ecology: Concepts and Applications
Ashley Waters

International Standard Book Number: 978-1-64740-035-4 (Hardback)

Cataloging-in-Publication Data

Marine ecology : concepts and applications / Ashley Waters.
p. cm.
Includes bibliographical references and index.
ISBN 978-1-64740-035-4
1. Marine ecology. 2. Aquatic ecology. 3. Marine habitats.
4. Marine ecosystem management. I. Waters, Ashley.
QH541.5.S3 M37 2020
577.7--dc23

Printed in China.

TABLE OF CONTENTS

PREFACE

This book has been written, keeping in view that students want more practical information. Thus, my aim has been to make it as comprehensive as possible for the readers. I would like to extend my thanks to my family and co-workers for their knowledge, support and encouragement all along.

The scientific study of marine-life populations, habitat and interactions among organisms and the surrounding environment is known as marine ecology. It includes the study of their abiotic as well as biotic factors. Marine ecosystems are characterized by a high salt content. Marine ecology is involved in the study of the influence of geology, meteorology, geography, chemistry, physics and pedology of marine ecosystems. Some of the different types of marine ecosystems which are studied within this field are salt marshes, lagoons, mangroves, coral reefs and intertidal zones. It also studies the impact of human activities such as agriculture, medical research, fisheries and forestry on these ecosystems. This book explores all the important aspects of marine ecology in the present day scenario. It is a complete source of knowledge on the present status of this important field. Coherent flow of topics, student-friendly language and extensive use of examples make this book an invaluable source of knowledge.

A brief description of the chapters is provided below for further understanding:

Chapter – Introduction

Marine ecology is the study of the Earth's aquatic ecosystem, also known as marine ecosystem, which is characterized by water consisting of a high salt content. This is an introductory chapter which will introduce briefly all the significant aspects of marine ecosystems and marine ecology.

Chapter – Types of Marine Ecosystems

Some of the different types of marine ecosystems are estuaries, salt marshes, oceans, intertidal zones, coral reefs, lagoons and mangroves. This chapter has been carefully written to provide an easy understanding of these diverse types of marine ecosystems as well as why they should be protected.

Chapter – Understanding Marine Life

Marine life consists of various plants, animals and other organisms that live in seas or oceans. It includes different types of marine microorganisms, marine fauna, marine flora, zooplanktons, bacterioplanktons, marine invertebrates, etc. This chapter has been carefully written to provide an easy understanding of these types of marine life.

Chapter – Impact of Human Activities on Marine Ecology

There are numerous human activities which have an adverse effect upon marine ecology. A few of these are deep sea mining, deepwater drilling and fish farming. They can have numerous consequences such as marine pollution and accumulation of marine debris. This chapter discusses in detail these human activities that negatively affect the marine ecology.

Chapter – Marine Conservation

Marine conservation is concerned with the study of marine plants, animal resources and ecosystem functions. It aims to protect and preserve marine ecosystems in oceans and seas by preventing the exploitation of these resources. This chapter discusses in detail the different aspects of marine conservation.

Ashley Waters

Introduction 1

- Marine Ecology
- Marine Ecosystem

Marine ecology is the study of the Earth's aquatic ecosystem, also known as marine ecosystem, which is characterized by water consisting of a high salt content. This is an introductory chapter which will introduce briefly all the significant aspects of marine ecosystems and marine ecology.

Marine Ecology

Marine Ecology is the scientific study of marine-life habitat, populations, and interactions among organisms and the surrounding environment including their abiotic (non-living physical and chemical factors that affect the ability of organisms to survive and reproduce) and biotic factors (living things or the materials that directly or indirectly affect an organism in its environment).

Marine ecology is a subset of the study of marine biology and includes observations at the biochemical, cellular, individual, and community levels as well as the study of marine ecosystems and the biosphere.

The study of marine ecology also includes the influence of geology, geography, meteorology, pedology, chemistry, and physics on marine environments. The impact of human activity such as medical research, development, agriculture, fisheries, and forestry is also studied under marine ecology. In some ways, marine ecology is more complex than the relatively straightforward study of a particular organism or environment because of the numerous interconnections, symbiotic relationships, and influence of many factors on a particular environment.

Marine Ecosystem

Marine ecosystem, comprises of living organisms in the ocean environment.

Marine waters cover two-thirds of the surface of the Earth. In some places the ocean is deeper than Mount Everest is high; for example, the Mariana Trench and the Tonga Trench in the western part of the Pacific Ocean reach depths in excess of 10,000 metres (32,800 feet). Within this ocean habitat live a wide variety of organisms that have evolved in response to various features of their environs.

The Earth formed approximately 4.5 billion years ago. As it cooled, water in the atmosphere condensed and the Earth was pummeled with torrential rains, which filled its great basins, forming seas. The primeval atmosphere and waters harboured the inorganic components hydrogen, methane, ammonia, and water. These substances are thought to have combined to form the first organic compounds when sparked by electrical discharges of lightning. Some of the earliest known organisms are cyanobacteria (formerly referred to as blue-green algae). Evidence of these early photosynthetic prokaryotes has been found in Australia in Precambrian marine sediments called stromatolites that are approximately 3 billion years old. Although the diversity of life-forms observed in modern oceans did not appear until much later, during the Precambrian (about 4.6 billion to 542 million years ago) many kinds of bacteria, algae, protozoa, and primitive metazoa evolved to exploit the early marine habitats of the world. During the Cambrian Period (about 542 million to 488 million years ago) a major radiation of life occurred in the oceans. Fossils of familiar organisms such as cnidaria (e.g., jellyfish), echinoderms (e.g., feather stars), precursors of the fishes (e.g., the protochordate Pikaia from the Burgess Shale of Canada), and other vertebrates are found in marine sediments of this age. The first fossil fishes are found in sediments from the Ordovician Period (about 488 million to 444 million years ago). Changes in the physical conditions of the ocean that are thought to have occurred in the Precambrian—an increase in the concentration of oxygen in seawater and a buildup of the ozone layer that reduced dangerous ultraviolet radiation—may have facilitated the increase and dispersal of living things.

The Marine Environment

Geography, Oceanography and Topography

The shape of the oceans and seas of the world has changed significantly throughout the past 600 million years. According to the theory of plate tectonics, the crust of the Earth is made up of many dynamic plates. There are two types of plates—oceanic and continental—which float on the surface of the Earth's mantle, diverging, converging, or sliding against one another. When two plates diverge, magma from the mantle wells up and cools, forming new crust; when convergence occurs, one plate descends—i.e., is subducted—below the other and crust is resorbed into the mantle. Examples of both processes are observed in the marine environment. Oceanic crust is created along oceanic ridges or rift areas, which are vast undersea mountain ranges such as the Mid-Atlantic Ridge. Excess crust is reabsorbed along subduction zones, which usually are marked by deep-sea trenches such as the Kuril Trench off the coast of Japan.

The shape of the ocean also is altered as sea levels change. During ice ages a higher proportion of the waters of the Earth is bound in the polar ice caps, resulting in a relatively low sea level. When the polar ice caps melt during interglacial periods, the sea level rises. These changes in sea level cause great changes in the distribution of marine environments such as coral reefs. For example, during the last Pleistocene Ice Age the Great Barrier Reef did not exist as it does today; the continental shelf on which the reef now is found was above the high-tide mark.

Marine organisms are not distributed evenly throughout the oceans. Variations in characteristics of the marine environment create different habitats and influence what types of organisms will inhabit them. The availability of light, water depth, proximity to land, and topographic complexity all affect marine habitats.

The availability of light affects which organisms can inhabit a certain area of a marine ecosystem. The greater the depth of the water, the less light can penetrate until below a certain depth there is no light whatsoever. This area of inky darkness, which occupies the great bulk of the ocean, is called the aphotic zone. The illuminated region above it is called the photic zone, within which are distinguished the euphotic and disphotic zones. The euphotic zone is the layer closer to the surface that receives enough light for photosynthesis to occur. Beneath lies the disphotic zone, which is illuminated but so poorly that rates of respiration exceed those of photosynthesis. The actual depth of these zones depends on local conditions of cloud cover, water turbidity, and ocean surface. In general, the euphotic zone can extend to depths of 80 to 100 metres and the disphotic zone to depths of 80 to 700 metres. Marine organisms are particularly abundant in the photic zone, especially the euphotic portion; however, many organisms inhabit the aphotic zone and migrate vertically to the photic zone every night. Other organisms, such as the tripod fish and some species of sea cucumbers and brittle stars, remain in darkness all their lives.

Marine environments can be characterized broadly as a water, or pelagic, environment and a bottom, or benthic, environment. Within the pelagic environment the waters are divided into the neritic province, which includes the water above the continental shelf, and the oceanic province, which includes all the open waters beyond the continental shelf. The high nutrient levels of the neritic province—resulting from dissolved materials in riverine runoff—distinguish this province from the oceanic. The upper portion of both the neritic and oceanic waters—the epipelagic zone—is where photosynthesis occurs; it is roughly equivalent to the photic zone. Below this zone lie the mesopelagic, ranging between 200 and 1,000 metres, the bathypelagic, from 1,000 to 4,000 metres, and the abyssalpelagic, which encompasses the deepest parts of the oceans from 4,000 metres to the recesses of the deep-sea trenches.

The benthic environment also is divided into different zones. The supralittoral is above the high-tide mark and is usually not under water. The intertidal, or littoral, zone ranges from the high-tide mark (the maximum elevation of the tide) to the shallow, offshore waters. The sublittoral is the environment beyond the low-tide mark and is often used to refer to substrata of the continental shelf, which reaches depths of between 150 and 300 metres. Sediments of the continental shelf that influence marine organisms generally originate from the land, particularly in the form of riverine runoff, and include clay, silt, and sand. Beyond the continental shelf is the bathyal zone, which occurs at depths of 150 to 4,000 metres and includes the descending continental slope and rise. The abyssal zone (between 4,000 and 6,000 metres) represents a substantial portion of the oceans. The deepest region of the oceans (greater than 6,000 metres) is the hadal zone of the deep-sea trenches. Sediments of the deep sea primarily originate from a rain of dead marine organisms and their wastes.

Physical and Chemical Properties of Seawater

The physical and chemical properties of seawater vary according to latitude, depth, nearness to land, and input of fresh water. Approximately 3.5 percent of seawater is composed of dissolved compounds, while the other 96.5 percent is pure water. The chemical composition of seawater reflects such processes as erosion of rock and sediments, volcanic activity, gas exchange with the atmosphere, the metabolic and breakdown products of organisms, and rain. In addition to carbon, the nutrients essential for living organisms include nitrogen and phosphorus, which are minor constituents of seawater and thus are often limiting factors in organic cycles of the ocean. Concentrations of phosphorus and nitrogen are generally low in the photic zone because they are rapidly taken up by marine organisms. The highest concentrations of these nutrients generally are found below 500 metres, a result of the decay of organisms. Other important elements include silicon and calcium (essential in the skeletons of many organisms such as fish and corals).

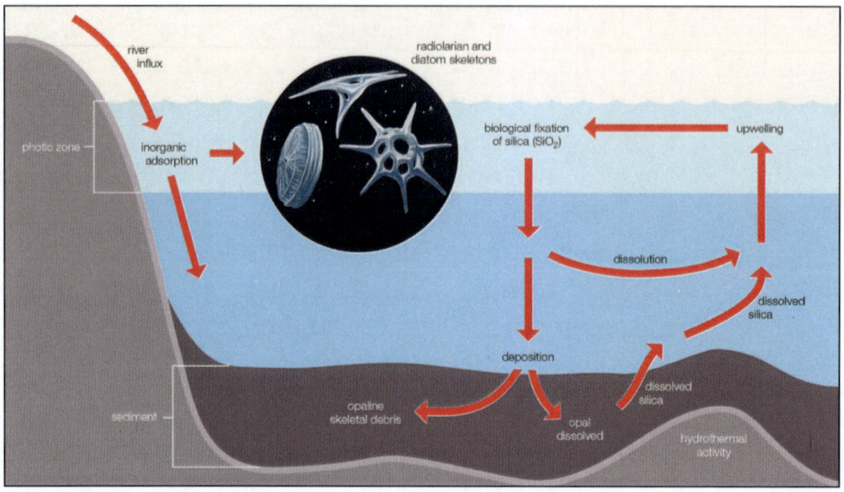

The chemical composition of the atmosphere also affects that of the ocean. For example, carbon dioxide is absorbed by the ocean and oxygen is released to the atmosphere through the activities of marine plants. The dumping of pollutants into the sea also can affect the chemical makeup of the ocean, contrary to earlier assumptions that, for example, toxins could be safely disposed of there.

The physical and chemical properties of seawater have a great effect on organisms, varying especially with the size of the creature. As an example, seawater is viscous to very small animals (less than 1 millimetre [0.039 inch] long) such as ciliates but not to large marine creatures such as tuna.

Marine organisms have evolved a wide variety of unique physiological and morphological features that allow them to live in the sea. Notothenid fishes in Antarctica are able to inhabit waters as cold as −2° C (28° F) because of proteins in their blood that act as antifreeze. Many organisms are able to achieve neutral buoyancy by secreting gas into internal chambers, as cephalopods do, or into swim bladders, as some fish do; other organisms use lipids, which are less dense than water, to achieve this effect. Some animals, especially those in the aphotic zone, generate light to attract prey. Animals in the disphotic zone such as hatchetfish produce light by means of organs called photophores to break up the silhouette of their bodies and avoid visual detection by predators. Many marine animals can detect vibrations or sound in the water over great distances by means of specialized organs. Certain fishes have lateral line systems, which they use to detect prey, and whales have a sound-producing organ called a melon with which they communicate. Tolerance to differences in salinity varies greatly: stenohaline organisms have a low tolerance to salinity changes, whereas euryhaline organisms, which are found in areas where river and sea meet (estuaries), are very tolerant of large changes in salinity. Euryhaline organisms are also very tolerant of changes in temperature. Animals that migrate between fresh water and salt water, such as salmon or eels, are capable of controlling their osmotic environment by active pumping or the retention of salts. Body architecture varies greatly in marine waters. The body shape of the cnidarian by-the-wind-sailor (Velella

velella)—an animal that lives on the surface of the water (pleuston) and sails with the assistance of a modified flotation chamber—contrasts sharply with the sleek, elongated shape of the barracuda.

Ocean Currents

The movements of ocean waters are influenced by numerous factors, including the rotation of the Earth (which is responsible for the Coriolis effect), atmospheric circulation patterns that influence surface waters, and temperature and salinity gradients between the tropics and the polar regions. The resultant patterns of circulation range from those that cover great areas, such as the North Subtropical Gyre, which follows a path thousands of kilometres long, to small-scale turbulences of less than one metre.

Marine organisms of all sizes are influenced by these patterns, which can determine the range of a species. For example, krill (Euphausia superba) are restricted to the Antarctic Circumpolar Current. Distribution patterns of both large and small pelagic organisms are affected as well. Mainstream currents such as the Gulf Stream and East Australian Current transport larvae great distances. As a result cold temperate coral reefs receive a tropical infusion when fish and invertebrate larvae from the tropics are relocated to high latitudes by these currents. The successful recruitment of eels to Europe depends on the strength of the Gulf Stream to transport them from spawning sites in the Caribbean. Areas where the ocean is affected by nearshore features, such as estuaries, or areas in which there is a vertical salinity gradient (halocline) often exhibit intense biological activity. In these environments, small organisms can become concentrated, providing a rich supply of food for other animals.

Marine Biota

Marine biota can be classified broadly into those organisms living in either the pelagic environment (plankton and nekton) or the benthic environment (benthos). Some organisms, however, are benthic in one stage of life and pelagic in another. Producers that synthesize organic molecules exist in both environments. Single-celled or multicelled plankton with photosynthetic pigments are the producers of the photic zone in the pelagic environment. Typical benthic producers are microalgae (e.g., diatoms), macroalgae (e.g., the kelp Macrocystis pyrifera), or sea grass (e.g., Zostera).

Plankton

Plankton are the numerous, primarily microscopic inhabitants of the pelagic environment. They are critical components of food chains in all marine environments because they provide nutrition for the nekton (e.g., crustaceans, fish, and squid) and benthos (e.g., sea squirts and sponges). They also exert a global effect on the biosphere because the balance of components of the Earth's atmosphere depends to a great extent on the photosynthetic activities of some plankton.

The term plankton is derived from the Greek planktos, meaning wandering or drifting, an apt description of the way most plankton spend their existence, floating with the ocean's currents. Not all plankton, however, are unable to control their movements, and many forms depend on self-directed motions for their survival.

Plankton range in size from tiny microbes (1 micrometre [0.000039 inch] or less) to jellyfish whose gelatinous bell can reach up to 2 metres in width and whose tentacles can extend over 15 metres.

However, most planktonic organisms, called plankters, are less than 1 millimetre (0.039 inch) long. These microbes thrive on nutrients in seawater and are often photosynthetic. The plankton include a wide variety of organisms such as algae, bacteria, protozoans, the larvae of some animals, and crustaceans. A large proportion of the plankton are protists—i.e., eukaryotic, predominantly single-celled organisms. Plankton can be broadly divided into phytoplankton, which are plants or plantlike protists; zooplankton, which are animals or animal-like protists; and microbes such as bacteria. Phytoplankton carry out photosynthesis and are the producers of the marine community; zooplankton are the heterotrophic consumers.

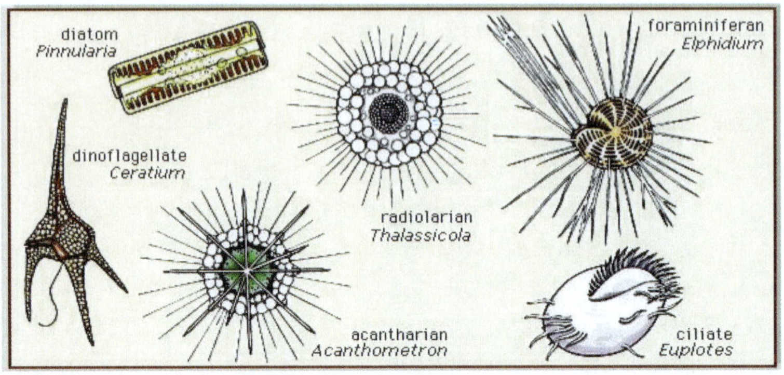

Representative plankton.

Diatoms and dinoflagellates (approximate range between 15 and 1,000 micrometres in length) are two highly diverse groups of photosynthetic protists that are important components of the plankton. Diatoms are the most abundant phytoplankton. While many dinoflagellates carry out photosynthesis, some also consume bacteria or algae. Other important groups of protists include flagellates, foraminiferans, radiolarians, acantharians, and ciliates. Many of these protists are important consumers and a food source for zooplankton.

Zooplankton, which are greater than 0.05 millimetre in size, are divided into two general categories: meroplankton, which spend only a part of their life cycle—usually the larval or juvenile stage—as plankton, and holoplankton, which exist as plankton all their lives. Many larval meroplankton in coastal, oceanic, and even freshwater environments (including sea urchins, intertidal snails, and crabs, lobsters, and fish) bear little or no resemblance to their adult forms. These larvae may exhibit features unique to the larval stage, such as the spectacular spiny armour on the larvae of certain crustaceans (e.g., Squilla), probably used to ward off predators.

Important holoplanktonic animals include such lobsterlike crustaceans as the copepods, cladocerans, and euphausids (krill), which are important components of the marine environment because they serve as food sources for fish and marine mammals. Gelatinous forms such as larvaceans, salps, and siphonophores graze on phytoplankton or other zooplankton. Some omnivorous zooplankton such as euphausids and some copepods consume both phytoplankton and zooplankton; their feeding behavior changes according to the availability and type of prey. The grazing and predatory activity of some zooplankton can be so intense that measurable reductions in phytoplankton or zooplankton abundance (or biomass) occur. For example, when jellyfish occur in high concentration in enclosed seas, they may consume such large numbers of fish larvae as to greatly reduce fish populations.

The jellylike plankton are numerous and predatory. They secure their prey with stinging cells (nematocysts) or sticky cells (colloblasts of comb jellies). Large numbers of the Portuguese man-of-war

(Physalia), with its conspicuous gas bladder, the by-the-wind-sailor (Velella velella), and the small blue disk-shaped Porpita porpita are propelled along the surface by the wind, and after strong onshore winds they may be found strewn on the beach. Beneath the surface, comb jellies often abound, as do siphonophores, salps, and scyphomedusae.

The pelagic environment was once thought to present few distinct habitats, in contrast to the array of niches within the benthic environment. Because of its apparent uniformity, the pelagic realm was understood to be distinguished simply by plankton of different sizes. Small-scale variations in the pelagic environment, however, have been discovered that affect biotic distributions. Living and dead matter form organic aggregates called marine snow to which members of the plankton community may adhere, producing patchiness in biotic distributions. Marine snow includes structures such as aggregates of cells and mucus as well as drifting macroalgae and other flotsam that range in size from 0.5 millimetre to 1 centimetre (although these aggregates can be as small as 0.05 millimetre and as large as 100 centimetres). Many types of microbes, phytoplankton, and zooplankton stick to marine snow, and some grazing copepods and predators will feed from the surface of these structures. Marine snow is extremely abundant at times, particularly after plankton blooms. Significant quantities of organic material from upper layers of the ocean may sink to the ocean floor as marine snow, providing an important source of food for bottom dwellers. Other structures that plankton respond to in the marine environment include aggregates of phytoplankton cells that form large rafts in tropical and temperate waters of the world (e.g., cells of Oscillatoria [Trichodesmium] erthraeus) and various types of seaweed (e.g., Sargassum, Phyllospora, Macrocystis) that detach from the seafloor and drift.

Nekton

Nekton are the active swimmers of the oceans and are often the best-known organisms of marine waters. Nekton are the top predators in most marine food chains. The distinction between nekton and plankton is not always sharp. Many large marine animals, such as marlin and tuna, spend the larval stage of their lives as plankton and their adult stage as large and active members of the nekton. Other organisms such as krill are referred to as both micronekton and macrozooplankton.

The vast majority of nekton are vertebrates (e.g., fishes, reptiles, and mammals), mollusks, and crustaceans. The most numerous group of nekton are the fishes, with approximately 16,000 species. Nekton are found at all depths and latitudes of marine waters. Whales, penguins, seals, and icefish abound in polar waters. Lantern fish (family Myctophidae) are common in the aphotic zone along with gulpers (Saccopharynx), whalefish (family Cetomimidae), seven-gilled sharks, and others. Nekton diversity is greatest in tropical waters, where in particular there are large numbers of fish species.

The largest animals on the Earth, the blue whales (Balaenoptera musculus), which grow to 25 to 30 metres long, are members of the nekton. These huge mammals and other baleen whales (order Mysticeti), which are distinguished by fine filtering plates in their mouths, feed on plankton and micronekton as do whale sharks (Rhinocodon typus), the largest fish in the world (usually 12 to 14 metres long, with some reaching 17 metres). The largest carnivores that consume large prey include the toothed whales (order Odontoceti—for example, the killer whales, Orcinus orca), great white sharks (Carcharodon carcharias), tiger sharks (Galeocerdo cuvier), black marlin (Makaira indica), bluefin tuna (Thunnus thynnus), and giant groupers (Epinephelus lanceolatus).

Nekton form the basis of important fisheries around the world. Vast schools of small anchovies, herring, and sardines generally account for one-quarter to one-third of the annual harvest from the ocean. Squid are also economically valuable nekton. Halibut, sole, and cod are demersal (i.e., bottom-dwelling) fish that are commercially important as food for humans. They are generally caught in continental shelf waters. Because pelagic nekton often abound in areas of upwelling where the waters are nutrient-rich, these regions also are major fishing areas.

Benthos

Organisms are abundant in surface sediments of the continental shelf and in deeper waters, with a great diversity found in or on sediments. In shallow waters, beds of seagrass provide a rich habitat for polychaete worms, crustaceans (e.g., amphipods), and fishes. On the surface of and within intertidal sediments most animal activities are influenced strongly by the state of the tide. On many sediments in the photic zone, however, the only photosynthetic organisms are microscopic benthic diatoms.

Benthic organisms can be classified according to size. The macrobenthos are those organisms larger than 1 millimetre. Those that eat organic material in sediments are called deposit feeders (e.g., holothurians, echinoids, gastropods), those that feed on the plankton above are the suspension feeders (e.g., bivalves, ophiuroids, crinoids), and those that consume other fauna in the benthic assemblage are predators (e.g., starfish, gastropods). Organisms between 0.1 and 1 millimetre constitute the meiobenthos. These larger microbes, which include foraminiferans, turbellarians, and polychaetes, frequently dominate benthic food chains, filling the roles of nutrient recycler, decomposer, primary producer, and predator. The microbenthos are those organisms smaller than 1 millimetre; they include diatoms, bacteria, and ciliates.

Organic matter is decomposed aerobically by bacteria near the surface of the sediment where oxygen is abundant. The consumption of oxygen at this level, however, deprives deeper layers of oxygen, and marine sediments below the surface layer are anaerobic. The thickness of the oxygenated layer varies according to grain size, which determines how permeable the sediment is to oxygen and the amount of organic matter it contains. As oxygen concentration diminishes, anaerobic processes come to dominate. The transition layer between oxygen-rich and oxygen-poor layers is called the redox discontinuity layer and appears as a gray layer above the black anaerobic layers. Organisms have evolved various ways of coping with the lack of oxygen. Some anaerobes release hydrogen sulfide, ammonia, and other toxic reduced ions through metabolic processes. The thiobiota, made up primarily of microorganisms, metabolize sulfur. Most organisms that live below the redox layer, however, have to create an aerobic environment for themselves. Burrowing animals generate a respiratory current along their burrow systems to oxygenate their dwelling places; the influx of oxygen must be constantly maintained because the surrounding anoxic layer quickly depletes the burrow of oxygen. Many bivalves (e.g., Mya arenaria) extend long siphons upward into oxygenated waters near the surface so that they can respire and feed while remaining sheltered from predation deep in the sediment. Many large mollusks use a muscular "foot" to dig with, and in some cases they use it to propel themselves away from predators such as starfish. The consequent "irrigation" of burrow systems can create oxygen and nutrient fluxes that stimulate the production of benthic producers (e.g., diatoms).

Not all benthic organisms live within the sediment; certain benthic assemblages live on a rocky substrate. Various phyla of algae—Rhodophyta (red), Chlorophyta (green), and Phaeophyta (brown)—are abundant and diverse in the photic zone on rocky substrata and are important producers. In intertidal regions algae are most abundant and largest near the low-tide mark. Ephemeral algae such as Ulva, Enteromorpha, and coralline algae cover a broad range of the intertidal. The mix of algae species found in any particular locale is dependent on latitude and also varies greatly according to wave exposure and the activity of grazers. For example, Ascophyllum spores cannot attach to rock in even a gentle ocean surge; as a result this plant is largely restricted to sheltered shores. The fastest-growing plant—adding as much as 1 metre per day to its length—is the giant kelp, Macrocystis pyrifera, which is found on subtidal rocky reefs. These plants, which may exceed 30 metres in length, characterize benthic habitats on many temperate reefs. Large laminarian and fucoid algae are also common on temperate rocky reefs, along with the encrusting (e.g., Lithothamnion) or short tufting forms (e.g., Pterocladia). Many algae on rocky reefs are harvested for food, fertilizer, and pharmaceuticals. Macroalgae are relatively rare on tropical reefs where corals abound, but Sargassum and a diverse assemblage of short filamentous and tufting algae are found, especially at the reef crest. Sessile and slow-moving invertebrates are common on reefs. In the intertidal and subtidal regions herbivorous gastropods and urchins abound and can have a great influence on the distribution of algae. Barnacles are common sessile animals in the intertidal. In the subtidal regions, sponges, ascidians, urchins, and anemones are particularly common where light levels drop and current speeds are high. Sessile assemblages of animals are often rich and diverse in caves and under boulders.

Reef-building coral polyps (Scleractinia) are organisms of the phylum Cnidaria that create a calcareous substrate upon which a diverse array of organisms live. Approximately 700 species of corals are found in the Pacific and Indian oceans and belong to genera such as Porites, Acropora, and Montipora. Some of the world's most complex ecosystems are found on coral reefs. Zooxanthellae are the photosynthetic, single-celled algae that live symbiotically within the tissue of corals and help to build the solid calcium carbonate matrix of the reef. Reef-building corals are found only in waters warmer than 18 °C; warm temperatures are necessary, along with high light intensity, for the coral-algae complex to secrete calcium carbonate. Many tropical islands are composed entirely of hundreds of metres of coral built atop volcanic rock.

Links between the Pelagic Environments and the Benthos

Considering the pelagic and benthic environments in isolation from each other should be done cautiously because the two are interlinked in many ways. For example, pelagic plankton are an important source of food for animals on soft or rocky bottoms. Suspension feeders such as anemones and barnacles filter living and dead particles from the surrounding water while detritus feeders graze on the accumulation of particulate material raining from the water column above. The molts of crustaceans, plankton feces, dead plankton, and marine snow all contribute to this rain of fallout from the pelagic environment to the ocean bottom. This fallout can be so intense in certain weather patterns—such as the El Niño condition—that benthic animals on soft bottoms are smothered and die. There also is variation in the rate of fallout of the plankton according to seasonal cycles of production. This variation can create seasonality in the abiotic zone where there is little or no variation in temperature or light. Plankton form marine sediments, and many types of fossilized protistan plankton, such as foraminiferans and coccoliths, are used to determine the age and origin of rocks.

Organisms of the Deep-Sea Vents

Producers were discovered in the aphotic zone when exploration of the deep sea by submarine became common in the 1970s. Deep-sea hydrothermal vents now are known to be relatively common in areas of tectonic activity (e.g., spreading ridges). The vents are a nonphotosynthetic source of organic carbon available to organisms. A diversity of deep-sea organisms including mussels, large bivalve clams, and vestimentiferan worms are supported by bacteria that oxidize sulfur (sulfide) and derive chemical energy from the reaction. These organisms are referred to as chemoautotrophic, or chemosynthetic, as opposed to photosynthetic, organisms. Many of the species in the vent fauna have developed symbiotic relationships with chemoautotrophic bacteria, and as a consequence the megafauna are principally responsible for the primary production in the vent assemblage. The situation is analogous to that found on coral reefs where individual coral polyps have symbiotic relationships with zooxanthellae. In addition to symbiotic bacteria there is a rich assemblage of free-living bacteria around vents. For example, Beggiatoas-like bacteria often form conspicuous weblike mats on any hard surface; these mats have been shown to have chemoautotrophic metabolism. Large numbers of brachyuran (e.g., Bythograea) and galatheid crabs, large sea anemones (e.g., Actinostola callasi), copepods, other plankton, and some fish—especially the eelpout Thermarces cerberus—are found in association with vents.

Hydrothermal mussels.

Patterns and Processes Influencing the Structure of Marine Assemblages

Distribution and Dispersal

The distribution patterns of marine organisms are influenced by physical and biological processes in both ecological time (tens of years) and geologic time (hundreds to millions of years). The shapes of the Earth's oceans have been influenced by plate tectonics, and as a consequence the distributions of fossil and extant marine organisms also have been affected. Vicariance theory argues that plate tectonics has a major role in determining biogeographic patterns. For example, Australia was once—90 million years ago—close to the South Pole and had few coral reefs. Since then Australia has been moving a few millimetres each year closer to the Equator. As a result of this movement and local oceanographic conditions, coral reef environments are extending ever so slowly southward. Dispersal may also have an important role in biogeographic patterns of abundance. The importance of dispersal varies greatly with local oceanographic features, such as the direction and intensity of currents and the biology of the organisms. Humans can also have an impact on patterns of distribution and the extinction of marine organisms. For example, fishing intensity in the Irish Sea was based on catch limits set for cod with no regard for the biology of other species. One consequence of this practice was that the local skate, which had a slow reproductive rate, was quickly fished to extinction.

A characteristic of many marine organisms is a bipartite life cycle, which can affect the dispersal of an organism. Most animals found on soft and hard substrata, such as lobsters, crabs, barnacles, fish, polychaete worms, and sea urchins, spend their larval phase in the plankton and in this phase are dispersed most widely. The length of the larval phase, which can vary from a few minutes to hundreds of days, has a major influence on dispersal. For example, wrasses of the genus Thalassoma have a long larval life, compared with many other types of reef fish, and populations of these fish are well dispersed to the reefs of isolated volcanic islands around the Pacific. The bipartite life cycle of algae also affects their dispersal, which occurs through algal spores. Although in general, spores disperse only a short distance from adult plants, limited swimming abilities—Macrocystis spores have flagella—and storms can disperse spores over greater distances.

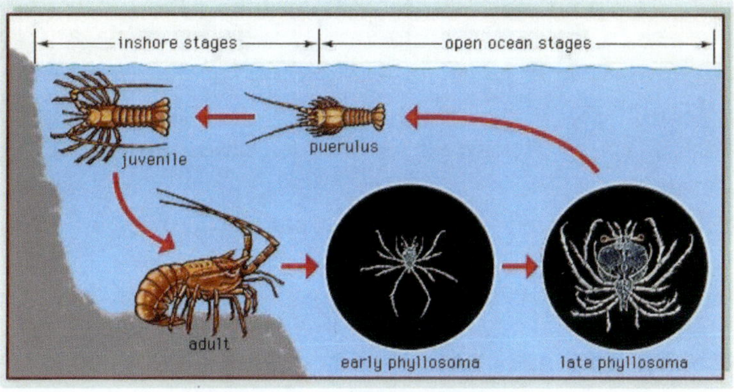

Life cycle of a palinurid lobster.

Migrations of Marine Organisms

The migrations of plankton and nekton throughout the water column in many parts of the world are well described. Diurnal vertical migrations are common. For example, some types of plankton, fish, and squid remain beneath the photic zone during the day, moving toward the surface after dusk and returning to the depths before dawn. It is generally argued that marine organisms migrate in response to light levels. This behavior may be advantageous because by spending the daylight hours in the dim light or darkness beneath the photic zone plankton can avoid predators that locate their prey visually. After the Sun has set, plankton can rise to the surface waters where food is more abundant and where they can feed safely under the cover of darkness.

Larval forms can facilitate their horizontal transport along different currents by migrating vertically. This is possible because currents can differ in direction according to depth (e.g., above and below haloclines and thermoclines), as is the case in estuaries.

In coastal waters many larger invertebrates (e.g., mysids, amphipods, and polychaete worms) leave the cover of algae and sediments to migrate into the water column at night. It is thought that these animals disperse to different habitats or find mates by swimming when visual predators find it hard to see them. In some cases only one sex will emerge at night, and often that sex is morphologically better suited for swimming.

Horizontal migrations of fish that span distances of hundreds of metres to tens of kilometres are common and generally related to patterns of feeding or reproduction. Tropical coral trout (Plectropomus species) remain dispersed over a reef for most of the year, but adults will aggregate at certain locations at the time of spawning. Transoceanic migrations (greater than 1,000 kilometres) are observed in a number of marine vertebrates, and these movements often relate to requirements of feeding and reproduction. Bluefin tuna (Thunnus thynnus) traverse the Atlantic Ocean in a single year; they spawn in the Caribbean, then swim to high latitudes of the Atlantic to feed on the rich supply of fish. Turtles and sharks also migrate great distances.

Fish that spend their lives in both marine and freshwater systems (diadromous animals) exhibit some of the most spectacular migratory behavior. Anadromous fishes (those that spend most of their lives in the sea but migrate to fresh water to spawn) such as Atlantic salmon (Salmo salar) also have unique migratory patterns. After spawning, the adults die. Newly hatched fish (alevin) emerge from spawned eggs and develop into young fry that move down rivers toward the sea. Juveniles (parr) grow into larger fish (smolt) that convene near the ocean. When the adult fish are ready to spawn, they return to the river in which they were born (natal river), using a variety of environmental cues, including the Earth's magnetic field, the Sun, and water chemistry. It is believed that the thyroid gland has a role in imprinting the water chemistry of the natal river on the fish. Freshwater eels such as the European eel (Anguilla anguilla) undertake great migrations from fresh water to spawn in the marine waters of the Sargasso Sea (catadromous migrations), where they die. Eel larvae, called leptocephalus larvae, drift back to Europe in the Gulf Stream.

Dynamics of Populations and Assemblages

A wide variety of processes influence the dynamics of marine populations of individual species and the composition of assemblages (e.g., collections of populations of different species that live in the same area). With the exception of marine mammals such as whales, fish that bear live young (e.g., embiotocid fish), and brooders (i.e., fauna that incubate their offspring until they emerge as larvae or juveniles), most marine organisms produce a large number of offspring of which few survive. Processes that affect the plankton can have a great influence on the numbers of young that survive to be recruited, or relocated, into adult populations. The survival of larvae may depend on the abundance of food at various times and in various places, the number of predators, and oceanographic features that retain larvae near suitable nursery areas. The number of organisms recruited to benthic and pelagic systems may ultimately determine the size of adult populations and therefore the relative abundance of species in marine assemblages. However, many processes can affect the survival of organisms after recruitment. Predators eat recruits, and mortality rates in prey species can vary with time and space, thus changing original population patterns established in recruitment.

Patterns of colonization and succession can have a significant impact on benthic assemblages. For example, when intertidal reefs are cleared experimentally, the assemblage of organisms that colonize the bare space often reflects the types of larvae available in local waters at the time. Tube worms may dominate if they establish themselves first; if they fail to do so, algal spores may colonize the shore first and inhibit the settlement of these worms. Competition between organisms may also play a role. Long-term data gathered over periods of more than 25 years from coral reefs have demonstrated that some corals (e.g., Acropora cytherea) competitively overgrow neighbouring corals. Physical disturbance from hurricanes destroys many corals, and during regrowth competitively inferior species can coexist with normally dominant species on the reef. Chemical defenses of sessile organisms also can deter the growth or cause increased mortality of organisms that settle on them. Ascidian larvae (e.g., Podoclavella) often avoid settling on sponges (e.g., Mycale); when this does occur, the larvae rarely reach adulthood.

Although the processes that determine species assemblages may be understood, variations occur in the composition of the plankton that make it difficult to predict patterns of colonization with great accuracy.

Biological Productivity

Primary productivity is the rate at which energy is converted by photosynthetic and chemosynthetic autotrophs to organic substances. The total amount of productivity in a region or system is gross primary productivity. A certain amount of organic material is used to sustain the life of producers; what remains is net productivity. Net marine primary productivity is the amount of organic material available to support the consumers (herbivores and carnivores) of the sea. The standing crop is the total biomass (weight) of vegetation. Most primary productivity is carried out by pelagic phytoplankton, not benthic plants.

Most primary producers require nitrogen and phosphorus, which are available in the ocean as nitrate, nitrite, ammonia, and phosphorus. The abundances of these molecules and the intensity and quality of light exert a major influence on rates of production. The two principal categories of producers (autotrophs) in the sea are pelagic phytoplankton and benthic microalgae and macroalgae. Benthic plants grow only on the fringe of the world's oceans and are estimated to produce only 5 to 10 percent of the total marine plant material in a year. Chemoautotrophs are the producers of the deep-sea vents.

Primary productivity is usually determined by measuring the uptake of carbon dioxide or the output of oxygen. Production rates are usually expressed as grams of organic carbon per unit area per unit time. The productivity of the entire ocean is estimated to be approximately 16 × 1010 tons of carbon per year, which is about eight times that of the land.

Pelagic Food Chain

Food chains in coastal waters of the world are generally regulated by nutrient concentrations. These concentrations determine the abundance of phytoplankton, which in turn provide food for the primary consumers, such as protozoa and zooplankton, that the higher-level consumers—fish, squid, and marine mammals—prey upon. It had been thought that phytoplankton in the 5- to 100-micrometre size range were responsible for most of the primary production in the sea and that grazers such as copepods controlled the numbers of phytoplankton. Data gathered since 1975, however, indicate that the system is much more complex than this. It is now thought that most primary production in marine waters of the world is accomplished by single-celled 0.5- to 10-micrometre phototrophs (bacteria and protists). Moreover, heterotrophic protists (phagotrophic protists) are now viewed as the dominant controllers of both bacteria and primary production in the sea. Current models of pelagic marine food chains picture complex interactions within a microbial food web. Larger metazoans are supported by the production of autotrophic and heterotrophic cells.

Upwelling

The most productive waters of the world are in regions of upwelling. Upwelling in coastal waters brings nutrients toward the surface. Phytoplankton reproduce rapidly in these conditions, and grazing zooplankton also multiply and provide abundant food supplies for nekton. Some of the world's richest fisheries are found in regions of upwelling—for example, the temperate waters off Peru and California. If upwelling fails, the effects on animals that depend on it can be disastrous. Fisheries also suffer at these times, as evidenced by the collapse of the Peruvian anchovy industry in the 1970s. The intensity and location of upwelling are influenced by changes in atmospheric circulation, as exemplified by the influence of El Niño conditions.

Seasonal Cycles of Production

Cycles of plankton production vary at different latitudes because seasonal patterns of light and temperature vary dramatically with latitude. In the extreme conditions at the poles, plankton populations crash during the constant darkness of winter and bloom in summer with long hours of light and the retreat of the ice field. In tropical waters, variation in sunlight and temperature is slight, nutrients are present in low concentrations, and planktonic assemblages do not undergo large fluctuations in abundance. There are, however, rapid cycles of reproduction and high rates of grazing and predation that result in a rapid turnover of plankton and a low standing crop. In temperate regions plankton abundance peaks in spring as temperature and the length and intensity of daylight increase. Moreover, seasonal winter storms usually mix the water column, creating a more even distribution of the nutrients, which facilitates the growth of phytoplankton. Peak zooplankton production generally lags behind that of phytoplankton, while the consumption of phytoplankton by zooplankton and phagotrophic protists is thought to reduce phytoplankton abundance. Secondary peaks in abundance occur in autumn. Seasonal peaks of some plankton are very conspicuous, and the composition of the plankton varies considerably. In spring and early summer many fish and invertebrates spawn and release eggs and larvae into the plankton, and, as a result, the meroplanktonic component of the plankton is higher at these times. General patterns of plankton abundance may be further influenced by local conditions. Heavy rainfall in coastal regions (especially areas in which monsoons prevail) can result in nutrient-rich turbid plumes (i.e., estuarine or riverine plumes) that extend into waters of the continental shelf. Changes in production, therefore, may depend on the season, the proximity to fresh water, and the timing and location of upwelling, currents, and patterns of reproduction.

Types of Marine Ecosystems

2

- **Estuary**
- **Salt Marsh**
- **Ocean**
- **Intertidal Zone**
- **Coral Reefs**
- **Lagoon**
- **Mangrooves**

Some of the different types of marine ecosystems are estuaries, salt marshes, oceans, intertidal zones, coral reefs, lagoons and mangroves. This chapter has been carefully written to provide an easy understanding of these diverse types of marine ecosystems as well as why they should be protected.

Estuary

Estuary is a partly enclosed coastal body of water in which river water is mixed with seawater. In a general sense, the estuarine environment is defined by salinity boundaries rather than by geographic boundaries.

Estuaries are places where rivers meet the sea and may be defined as areas where salt water is measurably diluted with fresh water. On average, estuaries are biologically more productive than either the adjacent river or the sea, because they have a special kind of water circulation that traps plant nutrients and stimulates primary production. Fresh water, being lighter than salt water, tends to form a distinct layer that floats at the surface of the estuary. At the boundary between fresh and salt water, there is a certain amount of mixing caused by the flow of fresh water over salt and by the ebb and flow of tides. Additional mixing may be caused from time to time by strong winds and by internal waves that are propagated along the interface between fresh and salt water.

Many coastal features that are designated by other names are in fact estuaries. For example, various coastal embayments, such as Chesapeake Bay and Galveston Bay in the United States, also are estuaries, because fresh and salt water undergo considerable mixing. Moreover, most of the world's submerged fjord systems (such as Scoresby Sund, Greenland) and large semi-enclosed tidal flat regions and coastal marshes (such as the Waddenzee area of The Netherlands) are estuaries. In addition, estuaries include river mouths (as in the case of the Mississippi River, Louisiana), structural basins (San Francisco Bay, California), and the bodies of water behind spits (Hurst Castle spit, England) and barrier beaches (Ninety Mile Beach, Australia). In the case of spits and barrier beaches, the definitions of lagoons and estuaries overlap.

Types

The four basic types of estuaries are (1) the salt wedge estuary, (2) the partially mixed (or slightly stratified) estuary, (3) the vertically homogeneous (or vertically mixed) estuary, and (4) the fjord (or highly stratified estuary).

Salt Wedge Estuaries

A salt wedge estuary has minimal mixing, and the salt water forms a wedge, thickest at the seaward end, tapering to a very thin layer at the landward limit. The penetration of this wedge changes with the flow of the river. During flood conditions the wedge will retreat, and during low flows it will extend farther upriver. The mouth of the Mississippi River in the United States is a classic example. The mixing at the boundary between fresh and salt water causes the surface layer to entrain salt water and become more saline as it moves toward the sea. To compensate for the entrained salt water, there is a slow movement of the salt water up the estuary at depth. Because bottom waters are rich in nutrients derived from decomposing plant and animal remains, this circulation has the effect of pumping nutrients into the estuary and stimulating biological production.

Organic and inorganic particles carried by rivers tend to flocculate (aggregate into a mass) and sediment out when they encounter salt water. They sink from the freshwater layer to the salt wedge and are carried upstream. When the organic matter decomposes, it adds still more nutrients to the estuary. The inorganic matter settles on the bottom and provides an enriched sediment for flowering plants adapted to salt water. Between the tide marks, mangrove forests flourish in tropical conditions, and salt marshes form in temperate and subarctic conditions. Below low tide, sea grasses form dense beds on muddy substrates. In areas of an estuary where water movement is vigorous enough to remove sediment, leaving a stony or rocky bottom, rooted plants are replaced by seaweeds. These have a special structure known as a holdfast, which attaches itself to any hard surface. Phytoplankton floating freely in the water benefit from the high level of nutrients, especially near the head of the estuary, and grow rapidly, providing food for the microscopic animals in the water column, the zooplankton. As this community is carried downstream in the surface waters, dead organisms and the fecal pellets of the animals sink toward the bottom and enter the salt wedge to be carried back to the head of the estuary. As they decompose, they add still more nutrients to the water.

Partially Mixed Estuaries

In a partially mixed estuary, the vigorous rise and fall of the tide generates strong turbulence and causes partial mixing between the fresh water above and the salt water below. Under these conditions the river flow entrains 10 to 20 or more times its own volume of salt water, and the compensatory landward flow of seawater near the bottom is correspondingly increased. The effect of Earth's rotation (Coriolis effect) is to cause the surface flow to be stronger on the right-hand side facing seaward in the Northern Hemisphere and on the opposite side in the Southern Hemisphere. The bottom flow is stronger on the opposite side of the estuary.

Vertically Homogeneous Estuaries

In a vertically homogeneous estuary the river flow is weak, and the tidal flow is strong. Consequently, no stratification occurs, and salt water may even move up the river channel. Salinity levels are nearly the same from top to bottom at any given place in such estuaries; however, the salinity is lowest where the river enters the estuary and highest near the sea.

Fjords

The fjord-type estuary was originally formed by a glacier and has a submerged ridge, or sill, near its mouth, composed of glacial deposits. It may be regarded as a partially mixed estuary in which the bottom has been replaced by a basin of undiluted seawater held in place by the sill. When entrainment in river flow causes a strong landward flow at the bottom, water rises over the sill and enters the estuary at intermediate depth, leaving the deep waters undisturbed. Only major intrusions of seawater caused by storms can displace the deep water. Fjords, because of their glacial origin, commonly have steep sides and very little shallow water. Hence, the development of salt marshes or sea grass beds is minimal, but seaweeds colonize the rocky shores.

Geology and Geomorphology

The geologic processes that form an estuary are extremely complex and varied, but it is clear that the existence of an estuary is largely dependent on the position of sea level relative to the freshwater discharge. If sea level were lowered, the estuarine zone would migrate seaward at the interface of the marine water and the edge of the newly exposed land area. Such migration has occurred as a consequence of Earth's several glaciations. For each glaciation, the primary source of moisture has been the oceans. Whenever sea level fell, the estuarine environment at the continental margin was forced to migrate seaward.

About 18,000 years ago the Wisconsin Glacial Stage attained its maximum, and glacial melting began. The seas rose, forcing the estuarine environment to migrate back up the continental shelf. During the period of lowered sea level, some rivers had become entrenched in the continental shelf and deepened their valleys, which were soon flooded by the rising marine waters, forming a typical drowned river estuary. In areas such as Norway and parts of the coast of British Columbia, Canada, valley glaciers had deepened river valleys. These narrow drowned glacial valleys became the modern fjord estuaries as sea level rose.

The geomorphology of an estuarine basin is usually developed by one of three agents: (1) fluvial or glacial erosion, (2) fluvial and marine deposition, or (3) tectonic activity. The last of these involves the downfaulting of a coastal area or the broad local subsidence of a stretch of coastline, as in the case of San Francisco Bay.

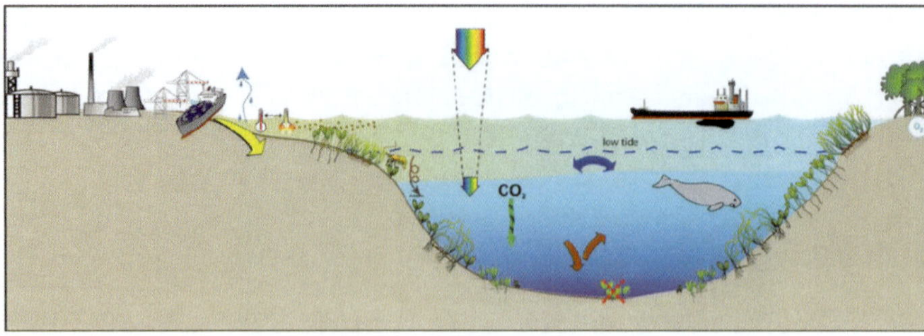

Salt Marsh

A salt marsh or saltmarsh, also known as a coastal salt marsh or a tidal marsh, is a coastal ecosystem in the upper coastal intertidal zone between land and open saltwater or brackish water that is regularly flooded by the tides. It is dominated by dense stands of salt-tolerant plants such as herbs, grasses, or low shrubs. These plants are terrestrial in origin and are essential to the stability of the salt marsh in trapping and binding sediments. Salt marshes play a large role in the aquatic food web and the delivery of nutrients to coastal waters. They also support terrestrial animals and provide coastal protection.

Salt marshes occur on low-energy shorelines in temperate and high-latitudes which can be stable, emerging, or submerging depending if the sedimentation is greater, equal to, or lower than relative sea level rise (subsidence rate plus sea level change), respectively. Commonly these shorelines consist of mud or sand flats (known also as tidal flats or abbreviated to mudflats) which are nourished with sediment from inflowing rivers and streams. These typically include sheltered environments such as embankments, estuaries and the leeward side of barrier islands and spits. In the tropics and sub-tropics they are replaced by mangroves; an area that differs from a salt marsh in that instead of herbaceous plants, they are dominated by salt-tolerant trees.

Most salt marshes have a low topography with low elevations but a vast wide area, making them hugely popular for human populations. Salt marshes are located among different landforms based on their physical and geomorphological settings. Such marsh landforms include deltaic marshes, estuarine, back-barrier, open coast, embayments and drowned-valley marshes. Deltaic marshes are associated with large rivers where many occur in Southern Europe such as the Camargue, France in the Rhone delta or the Ebro delta in Spain. They are also extensive within the rivers of the Mississippi Delta in the United States. In New Zealand, most salt marshes occur at the head of estuaries in areas where there is little wave action and high sedimentation. Such marshes are located in Awhitu Regional Park in Auckland, the Manawatu Estuary, and the Avon-Heathcote Estuary in Christchurch. Back-barrier marshes are sensitive to the reshaping of barriers in the landward side of which they have been formed. They are common along much of the eastern coast of the United States and the Frisian Islands. Large, shallow coastal embayments can hold salt marshes with examples including Morecambe Bay and Portsmouth in Britain and the Bay of Fundy in North America.

Salt marshes are sometimes included in lagoons, and the difference is not very marked; the Venetian Lagoon in Italy, for example, is made up of these sorts of animals and or living organisms belonging to this ecosystem. They have a big impact on the biodiversity of the area. Salt marsh ecology involves complex food webs which include primary producers (vascular plants, macroalgae, diatoms, epiphytes, and phytoplankton), primary consumers (zooplankton, macrozoa, molluscs, insects), and secondary consumers.

The low physical energy and high grasses provide a refuge for animals. Many marine fish use salt marshes as nursery grounds for their young before they move to open waters. Birds may raise their young among the high grasses, because the marsh provides both sanctuary from predators and abundant food sources which include fish trapped in pools, insects, shellfish, and worms.

An estuarine salt marsh along the Heathcote River, Christchurch, New Zealand.

Salt marsh on Sapelo Island, Georgia, USA.

Worldwide Occurrence

Saltmarshes across 99 countries (essentially worldwide) were mapped by Mcowen et al. 2017. A total of 5,495,089 hectares of mapped saltmarsh across 43 countries and territories are represented in a Geographic Information Systems polygon shapefile. This estimate is at the relatively low end of previous estimates (2.2–40 Mha). The most extensive saltmarshes worldwide are found outside the tropics, notably including the low-lying, ice-free coasts, bays and estuaries of the North Atlantic which are well represented in their global polygon dataset.

Formation

The formation begins as tidal flats gain elevation relative to sea level by sediment accretion, and subsequently the rate and duration of tidal flooding decreases so that vegetation can colonize on the exposed surface. The arrival of propagules of pioneer species such as seeds or rhizome portions are combined with the development of suitable conditions for their germination and establishment in the process of colonisation. When rivers and streams arrive at the low gradient of the tidal flats, the discharge rate reduces and suspended sediment settles onto the tidal flat surface, helped by the backwater effect of the rising tide. Mats of filamentous blue-green algae can fix silt and clay sized sediment particles to their sticky sheaths on contact which can also increase the erosion resistance of the sediments. This assists the process of sediment accretion to allow colonising species (e.g., Salicornia spp.) to grow. These species retain sediment washed in from the rising tide around their stems and leaves and form low muddy mounds which eventually coalesce to form depositional terraces, whose upward growth is aided by a sub-surface root network which binds the sediment. Once vegetation is established on depositional terraces further sediment trapping and accretion can allow rapid upward growth of the marsh surface such that there is an associated rapid decrease in the depth and duration of tidal flooding. As a result, competitive species that prefer higher elevations relative to sea level can inhabit the area and often a succession of plant communities develops.

Tidal Flooding and Vegetation Zonation

Coastal salt marshes can be distinguished from terrestrial habitats by the daily tidal flow that occurs and continuously floods the area. It is an important process in delivering sediments, nutrients and plant water supply to the marsh. At higher elevations in the upper marsh zone, there is much less tidal inflow, resulting in lower salinity levels. Soil salinity in the lower marsh zone is fairly constant due to everyday annual tidal flow. However, in the upper marsh, variability in salinity is shown as a result of less frequent flooding and climate variations. Rainfall can reduce salinity and evapotranspiration can increase levels during dry periods. As a result, there are microhabitats populated by different species of flora and fauna dependant on their physiological abilities. The flora of a salt marsh is differentiated into levels according to the plants' individual tolerance of salinity and water table levels. Vegetation found at the water must be able to survive high salt concentrations, periodical submersion, and a certain amount of water movement, while plants further inland in the marsh can sometimes experience dry, low-nutrient conditions. It has been found that the upper marsh zones limit species through competition and the lack of habitat protection, while lower marsh zones are determined through the ability of plants to tolerate physiological stresses such as salinity, water submergence and low oxygen levels.

The New England salt marsh is subject to strong tidal influences and shows distinct patterns of zonation. In low marsh areas with high tidal flooding, a monoculture of the smooth cordgrass, Spartina alterniflora dominate, then heading landwards, zones of the salt hay, Spartina patens, black rush, Juncus gerardii and the shrub Iva frutescens are seen respectively. These species all have different tolerances that make the different zones along the marsh best suited for each individual.

Plant species diversity is relatively low, since the flora must be tolerant of salt, complete or partial submersion, and anoxic mud substrate. The most common salt marsh plants are glassworts (Salicornia spp.) and the cordgrass (Spartina spp.), which have worldwide distribution. They are often the first plants to take hold in a mudflat and begin its ecological succession into a salt marsh. Their shoots lift the main flow of the tide above the mud surface while their roots spread into the substrate and stabilize the sticky mud and carry oxygen into it so that other plants can establish themselves as well. Plants such as sea lavenders (Limonium spp.), plantains (Plantago spp.), and varied sedges and rushes grow once the mud has been vegetated by the pioneer species.

Salt marshes are quite photosynthetically active and are extremely productive habitats. They serve as depositories for a large amount of organic matter and are full of decomposition, which feeds a broad food chain of organisms from bacteria to mammals. Many of the halophytic plants such as cordgrass are not grazed at all by higher animals but die off and decompose to become food for micro-organisms, which in turn become food for fish and birds.

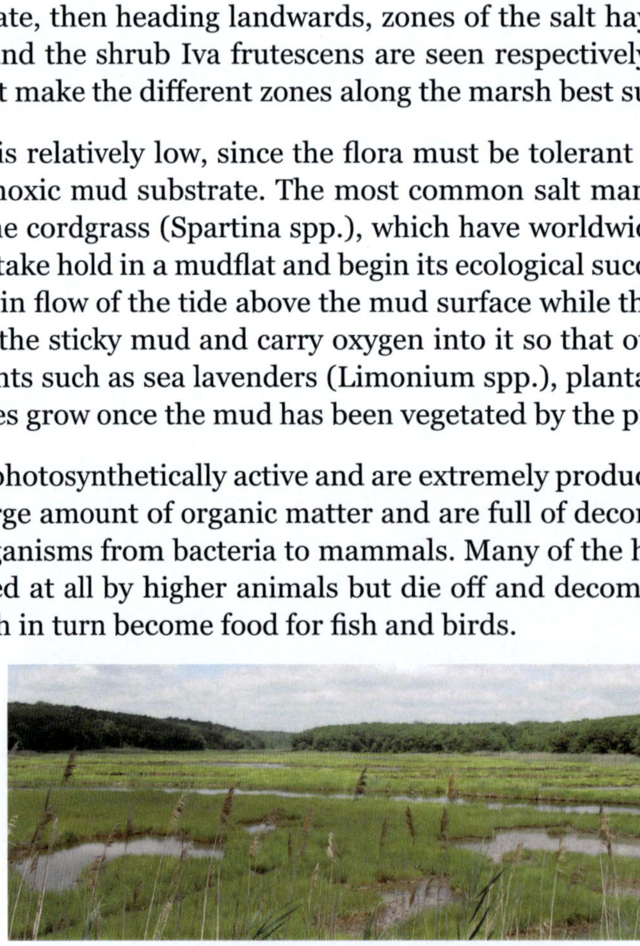

An Atlantic coastal salt marsh in Connecticut.

High marsh in the Marine Park Salt Marsh Nature Center in Brooklyn, New York.

Sediment Trapping, Accretion and the Role of Tidal Creeks

The factors and processes that influence the rate and spatial distribution of sediment accretion within the salt marsh are numerous. Sediment deposition can occur when marsh species provide a surface for the sediment to adhere to, followed by deposition onto the marsh surface when the sediment flakes off at low tide. The amount of sediment adhering to salt marsh species is dependent on the type of marsh species, the proximity of the species to the sediment supply, the amount of plant biomass, and the elevation of the species. For example, in a study of the Eastern Chongming Island and Jiuduansha Island tidal marshes at the mouth of the Yangtze River, China, the amount of sediment adhering to the species Spartina alterniflora, Phragmites australis, and Scirpus mariqueter decreased with distance from the highest levels of suspended sediment concentrations (found at the marsh edge bordering tidal creeks or the mudflats); decreased with those species at the highest elevations, which experienced the lowest frequency and depth of tidal inundations; and increased with increasing plant biomass. Spartina alterniflora, which had the most sediment adhering to it, may contribute >10% of the total marsh surface sediment accretion by this process.

Salt marsh species also facilitate sediment accretion by decreasing current velocities and encouraging sediment to settle out of suspension. Current velocities can be reduced as the stems of tall marsh species induce hydraulic drag, with the effect of minimising re-suspension of sediment and encouraging deposition. Measured concentrations of suspended sediment in the water column have been shown to decrease from the open water or tidal creeks adjacent to the marsh edge, to the marsh interior, probably as a result of direct settling to the marsh surface by the influence of the marsh canopy.

Inundation and sediment deposition on the marsh surface is also assisted by tidal creeks which are a common feature of salt marshes. Their typically dendritic and meandering forms provide avenues for the tide to rise and flood the marsh surface, as well as to drain water, and they may facilitate higher amounts of sediment deposition than salt marsh bordering open ocean. Sediment deposition is correlated with sediment size: coarser sediments will deposit at higher elevations (closer to the creek) than finer sediments (further from the creek). Sediment size is also often correlated with particular trace metals, and can thus tidal creeks can affect metal distributions and concentrations in salt marshes, in turn affecting the biota. Salt marshes do not however require tidal creeks to facilitate sediment flux over their surface although salt marshes with this morphology seem to be rarely studied.

The elevation of marsh species is important; those species at lower elevations experience longer and more frequent tidal floods and therefore have the opportunity for more sediment deposition to occur. Species at higher elevations can benefit from a greater chance of inundation at the highest tides when increased water depths and marsh surface flows can penetrate into the marsh interior.

Bloody Marsh in Georgia, USA.

Human Impacts

The coast is a highly attractive natural feature to humans through its beauty, resources, and accessibility. As of 2002, over half of the world's population was estimated to being living within 60 km of the coastal shoreline, making coastlines highly vulnerable to human impacts from daily activities that put pressure on these surrounding natural environments. In the past, salt marshes were perceived as coastal 'wastelands,' causing considerable loss and change of these ecosystems through land reclamation for agriculture, urban development, salt production and recreation. The indirect effects of human activities such as nitrogen loading also play a major role in the salt marsh area. Salt marshes can suffer from dieback in the high marsh and die-off in the low marsh.

Land Reclamation

Reclamation of land for agriculture by converting marshland to upland was historically a common practice. Dikes were often built to allow for this shift in land change and to provide flood protection further inland. In recent times intertidal flats have also been reclaimed. For centuries, livestock such as sheep and cattle grazed on the highly fertile salt marsh land. Land reclamation for agriculture has resulted in many changes such as shifts in vegetation structure, sedimentation, salinity, water flow, biodiversity loss and high nutrient inputs. There have been many attempts made to eradicate these problems for example, in New Zealand, the cordgrass Spartina anglica was introduced from England into the Manawatu River mouth in 1913 to try and reclaim the estuary land for farming. A shift in structure from bare tidal flat to pastureland resulted from increased sedimentation and the cordgrass extended out into other estuaries around New Zealand. Native plants and animals struggled to survive as non-natives out competed them. Efforts are now being made to remove these cordgrass species, as the damages are slowly being recognized.

In the Blyth estuary in Suffolk in eastern England, the mid-estuary reclamations (Angel and Bulcamp marshes) that were abandoned in the 1940s have been replaced by tidal flats with compacted soils from agricultural use overlain with a thin veneer of mud. Little vegetation colonisation has occurred in the last 60–75 years and has been attributed to a combination of surface elevations too low for pioneer species to develop, and poor drainage from the compacted agricultural soils acting as an aquaclude. Terrestrial soils of this nature need to adjust from fresh to saline interstitial water by a change in the chemistry and the structure of the soil, accompanied with fresh deposition of estuarine sediment, before salt marsh vegetation can establish. The vegetation structure, species richness, and plant community composition of salt marshes naturally regenerated on reclaimed agricultural land can be compared to adjacent reference salt marshes to assess the success of marsh regeneration.

Spartina alterniflora (Saltmarsh Cordgrass). Native to the eastern seaboard of the United States. Considered a noxious weed in the Pacific Northwest.

Upstream Agriculture

Cultivation of land upstream from the salt marsh can introduce increased silt inputs and raise the rate of primary sediment accretion on the tidal flats, so that pioneer species can spread further onto the flats and grow rapidly upwards out of the level of tidal inundation. As a result, marsh surfaces in this regime may have an extensive cliff at their seaward edge. At the Plum Island estuary, Massachusetts (U.S.A), stratigraphic cores revealed that during the 18th and 19th century the marsh prograded over subtidal and mudflat environments to increase in area from 6 km2 to 9 km^2 after European settlers deforested the land uptream and increased the rate of sediment supply.

Urban Development and Nitrogen Loading

The conversion of marshland to upland for agriculture has in the past century been overshadowed by conversion for urban development. Coastal cities worldwide have encroached onto former salt marshes and in the U.S. the growth of cities looked to salt marshes for waste disposal sites. Estuarine pollution from organic, inorganic, and toxic substances from urban development or industrialisation is a worldwide problem and the sediment in salt marshes may entrain this pollution with toxic effects on floral and faunal species. Urban development of salt marshes has slowed since about 1970 owing to growing awareness by environmental groups that they provide beneficial ecosystem services. They are highly productive ecosystems, and when net productivity is measured in $g\ m^{-2}\ yr^{-1}$ they are equalled only by tropical rainforests. Additionally, they can help reduce wave erosion on sea walls designed to protect low-lying areas of land from wave erosion.

De-naturalisation of the landward boundaries of salt marshes from urban or industrial enchroachment can have negative effects. In the Avon-Heathcote estuary/Ihutai, New Zealand, species abundance and the physical properties of the surrounding margins were strongly linked, and the majority of salt marsh was found to be living along areas with natural margins in the Avon and Heathcote river outlets; conversely, artificial margins contained little marsh vegetation and restricted landward retreat. The remaining marshes surrounding these urban areas are also under immense pressure from the human population as human-induced nitrogen enrichment enters these habitats. Nitrogen loading through human-use indirectly affects salt marshes causing shifts in vegetation structure and the invasion of non-native species.

Human impacts such as sewage, urban run-off, agricultural and industrial wastes are running into the marshes from nearby sources. Salt marshes are nitrogen limited and with an increasing level of nutrients entering the system from anthropogenic effects, the plant species associated with salt marshes are being restructured through change in competition. For example, the New England salt marsh is experiencing a shift in vegetation structure where S. alterniflora is spreading from the lower marsh where it predominately resides up into the upper marsh zone. Additionally, in the same marshes, the reed Phragmites australis has been invading the area expanding to lower marshes and becoming a dominant species. P. australis is an aggressive halophyte that can invade disturbed areas in large numbers outcompeting native plants. This loss in biodiversity is not only seen in flora assemblages but also in many animals such as insects and birds as their habitat and food resources are altered.

Chaetomorpha linum is a common marine algae found in the salt marsh.

Sea Level Rise

Due to the melting of Arctic sea ice and thermal expansion of the oceans, as a result of global warming, sea levels have begun to rise. As with all coastlines, this rise in water levels is predicted to negatively affect salt marshes, by flooding and eroding them. The sea level rise causes more open water zones within the salt marsh. These zones cause erosion along their edges, further eroding the marsh into open water until the whole marsh disintegrates.

While salt marshes are susceptible to threats concerning sea level rise, they are also an extremely dynamic coastal ecosystem. Salt marshes may in fact have the capability to keep pace with a rising sea level, by 2100, mean sea level could see increases between 0.6m to 1.1m. Marshes are susceptible to both erosion and accretion, which play a role in a what is called a bio-geomorphic feedback. Salt marsh vegetation captures sediment to stay in the system which in turn allows for the plants to grow better and thus the plants are better at trapping sediment and accumulate more organic matter. This positive feedback loop potentially allows for salt marsh bed level rates to keep pace with rising sea level rates. However, this feedback is also dependent on other factors like productivity of the vegetation, sediment supply, land subsidence, biomass accumulation, and magnitude and frequency of storms. In a study published by U.S.N. Best in 2018, they found that bioaccumulation was the number one factor in a salt marsh's ability to keep up with SLR rates. The salt marsh's resilience depends upon its increase in bed level rate being greater than that of sea levels increasing rate, otherwise the marsh will be overtaken and drowned.

Biomass accumulation can be measured in the form of above-ground organic biomass accumulation, and below-ground inorganic accumulation by means of sediment trapping and sediment settling from suspension. Salt marsh vegetation helps to increase sediment settling because it slows current velocities, disrupts turbulent eddies, and helps to dissipate wave energy. Marsh plant species are known for the tolerance of increased salt exposure due to the common inundation of marshlands. These types of plants are called halophytes. Halophytes are a crucial part of salt marsh biodiversity and their potential to adjust to elevated sea levels. With elevated sea levels, salt marsh vegetation would likely be more exposed to more frequent inundation rates and they must be adaptable or tolerant of the consequential increased salinity levels and anaerobic conditions. There is a common elevation (above the sea level) limit for these plants to survive, where anywhere below the optimal line would lead to anoxic soils due to constant submergence and too high above this line would mean harmful soil salinity levels due to the high rate of evapotranspiration as a result of decreased submergence. Along with the vertical accretion of sediment and biomass, the accommodation space for marsh land growth must also be considered. Accommodation space is the land available for additional sediments to accumulate and marsh vegetation to colonize laterally. This lateral accommodation space is often limited by anthropogenic structures such as coastal roads, sea walls, and other forms of development of coastal lands. That across a range of sea level rise rates, marshlands with high plant productivity were resistant against sea level rises but all reached a pinnacle point where accommodation space was necessary for continued survival. The presence of accommodation space allows for new mid/high habitat to form, and for marshes to escape complete inundation.

Mosquito Control

Earlier in the 20th century, it was believed that draining salt marshes would help reduce mosquito populations. In many locations, particularly in the northeastern United States, residents and local and state agencies dug straight-lined ditches deep into the marsh flats. The end result, however, was a depletion of killifish habitat. The killifish is a mosquito predator, so the loss of habitat actually led to higher mosquito populations, and adversely affected wading birds that preyed on the killifish. These ditches can still be seen, despite some efforts to refill the ditches.

Crab Herbivory and Bioturbation

Increased nitrogen uptake by marsh species into their leaves can prompt greater rates of length-specific leaf growth, and increase the herbivory rates of crabs. The burrowing crab Neohelice granulata frequents SW Atlantic salt marshes where high density populations can be found among populations of the marsh species Spartina densiflora and Sarcocornia perennis. In Mar Chiquita lagoon, north of Mar del Plata, Argentina, Neohelice granulata herbivory increased as a likely response to the increased nutrient value of the leaves of fertilised Spartina densiflora plots, compared to non-fertilised plots. Regardless of whether the plots were fertilised or not, grazing by Neohelice granulata also reduced the length specific leaf growth rates of the leaves in summer, while increasing their length-specific senescence rates. This may have been assisted by the increased fungal effectiveness on the wounds left by the crabs.

The salt marshes of Cape Cod, Massachusetts (U.S.A), are experiencing creek bank die-offs of Spartina spp. (cordgrass) that has been attributed to herbivory by the crab Sesarma reticulatum. At 12 surveyed Cape Cod salt marsh sites, 10% – 90% of creek banks experienced die-off of cordgrass in association with a highly denuded substrate and high density of crab burrows. Populations of

Sesarma reticulatum are increasing, possibly as a result of the degradation of the coastal food web in the region. The bare areas left by the intense grazing of cordgrass by Sesarma reticulatum at Cape Cod are suitable for occupation by another burrowing crab, Uca pugnax, which are not known to consume live macrophytes. The intense bioturbation of salt marsh sediments from this crab's burrowing activity has been shown to dramatically reduce the success of Spartina alterniflora and Suaeda maritima seed germination and established seedling survival, either by burial or exposure of seeds, or uprooting or burial of established seedlings. However, bioturbation by crabs may also have a positive effect. In New Zealand, the tunnelling mud crab Helice crassa has been given the stately name of an 'ecosystem engineer' for its ability to construct new habitats and alter the access of nutrients to other species. Their burrows provide an avenue for the transport of dissolved oxygen in the burrow water through the oxic sediment of the burrow walls and into the surrounding anoxic sediment, which creates the perfect habitat for special nitrogen cycling bacteria. These nitrate reducing (denitrifying) bacteria quickly consume the dissolved oxygen entering into the burrow walls to create the oxic mud layer that is thinner than that at the mud surface. This allows a more direct diffusion path for the export of nitrogen (in the form of gaseous nitrogen (N_2)) into the flushing tidal water.

Restoration and Management

The perception of bay salt marshes as a coastal 'wasteland' has since changed, acknowledging that they are one of the most biologically productive habitats on earth, rivalling tropical rainforests. Salt marshes are ecologically important providing habitats for native migratory fish and acting as sheltered feeding and nursery grounds. They are now protected by legislation in many countries to look after these ecologically important habitats. In the United States and Europe, they are now accorded to a high level of protection by the Clean Water Act and the Habitats Directive respectively. With the impacts of this habitat and its importance now realised, a growing interest in restoring salt marshes, through managed retreat or the reclamation of land has been established. However, many Asian countries such as China are still to recognise the value of marshlands. With their ever-growing populations and intense development along the coast, the value of salt marshes tends to be ignored and the land continues to be reclaimed.

Bakker et al. suggests two options available for restoring salt marshes. The first is to abandon all human interference and leave the salt marsh to complete its natural development. These types of restoration projects are often unsuccessful as vegetation tends to struggle to revert to its original structure and the natural tidal cycles are shifted due to land changes. The second option suggested by Bakker et al. is to restore the destroyed habitat into its natural state either at the original site or as a replacement at a different site. Under natural conditions, recovery can take 2–10 years or even longer depending on the nature and degree of the disturbance and the relative maturity of the marsh involved. Marshes in their pioneer stages of development will recover more rapidly than mature marshes as they are often first to colonize the land. It is important to note, that restoration can often be sped up through the replanting of native vegetation.

This last approach is often the most practiced and generally more successful than allowing the area to naturally recover on its own. The salt marshes in the state of Connecticut in the United States have long been an area lost to fill and dredging. As of 1969, the Tidal Wetland Act was introduced that ceased this practice, but despite the introduction of the act, the system was still degrading due to alterations in tidal flow. One area in Connecticut is the marshes on Barn Island. These marshes were diked then impounded with salt and brackish marsh during 1946–1966. As a result, the marsh shifted to a freshwater state and became dominated by the invasive species P. australis, Typha angustifolia and T. latifolia that have little ecological connection to the area.

By 1980, a restoration programme was put in place that has now been running for over 20 years. This programme has aimed to reconnect the marshes by returning tidal flow along with the ecological functions and characteristics of the marshes back to their original state. In the case of Barn Island, declines in the invasive species have initiated, re-establishing the tidal-marsh vegetation along with animal species such as fish and insects. This example highlights that considerable time and effort is needed to effectively restore salt marsh systems. Times in marsh recovery can depend on the development stage of the marsh; type and extent of the disturbance; geographical location; and the environmental and physiological stress factors to the marsh-associated flora and fauna.

Although much effort has gone into restoring salt marshes worldwide, further research is needed. There are many setbacks and problems associated with marsh restoration that requires careful long-term monitoring. Information on all components of the salt marsh ecosystem should be understood and monitored from sedimentation, nutrient, and tidal influences, to behavior patterns and tolerances of both flora and fauna species. Once a better understanding of these processes is acquired, and not just locally, but over a global scale, then more sound and practical management and restoration efforts can be implemented to preserve these valuable marshes and restore them to their original state.

While humans are situated along coastlines, there will always be the possibility of human-induced disturbances despite the number of restoration efforts we plan to implement. Dredging, pipelines for offshore petroleum resources, highway construction, accidental toxic spills or just plain carelessness are examples that will for some time now and into the future be the major influences of salt marsh degradation.

Glasswort (Salicornia spp.) a species endemic to the high marsh zone.

Common reed (Phragmites australis) an invasive species in degraded marshes in the northeastern United States.

In addition to restoring and managing salt marsh systems based on scientific principles, the opportunity should be taken to educate public audiences of their importance biologically and their purpose as serving as a ntural buffer for flood protection. Because salt marshes are often located next to urban areas, they are likely to receive more visitors than remote wetlands. By physically seeing the marsh, people are more likely to take notice and be more aware of the environment around them. An example of public involvement occurred at the Famosa Slough State Marine Conservation Area in San Diego, where a "friends" group worked for over a decade in trying to prevent the area from being developed. Eventually, the 5 hectare site was bought by the City and the group worked together to restore the area. The project involved removing invasive species and replanting with natives, along with public talks to other locals, frequent bird walks and clean-up events.

Ocean

An ocean is a body of water that composes much of a planet's hydrosphere.

When viewed from space, the predominance of Earth's oceans is readily apparent. The oceans and their marginal seas cover nearly 71 percent of Earth's surface, with an average depth of 3,688 metres (12,100 feet). The exposed land occupies the remaining 29 percent of the planetary surface and has a mean elevation of about 840 metres (approximately 2,755 feet). Actually, all the elevated land could be hidden under the oceans and Earth reduced to a smooth sphere that would be completely covered by a continuous layer of seawater more than 2,600 metres (8,530 feet) deep. This is known as the sphere depth of the oceans and serves to underscore the abundance of water on Earth's surface.

Earth: A composite image of Earth captured by instruments aboard NASA's.

Earth is unique in the solar system because of its distance from the Sun and its period of rotation. These combine to subject Earth to a solar radiation level that maintains the planet at a mean surface temperature of about 14–15 °C (57.2–59 °F). Mean surface temperature varies little over annual and night-day cycles. This mean temperature allows water to exist on Earth in all three of its phases—solid, liquid, and gaseous. No other planet in the solar system has this feature. The liquid phase predominates on Earth. By volume, 97.957 percent of the water on the planet exists as oceanic water and associated sea ice. The gaseous phase and droplet water in the atmosphere constitute 0.001 percent. Fresh water in lakes and streams makes up 0.036 percent, while groundwater is 10 times more abundant at 0.365 percent. Glaciers and ice caps constitute 1.641 percent of Earth's total water volume.

Each of the above is considered to be a reservoir of water. Water continuously circulates between these reservoirs in what is called the hydrologic cycle, which is driven by energy from the Sun. Evaporation, precipitation, movement of the atmosphere, and the downhill flow of river water, glaciers, and groundwater keep water in motion between the reservoirs and maintain the hydrologic cycle.

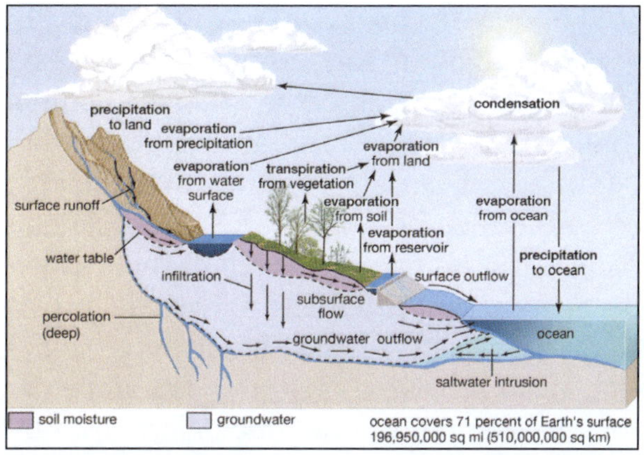

In the hydrologic cycle, water is transferred between the land surface, the ocean, and the atmosphere.

The large range of volumes in these reservoirs and the rates at which water cycles between them combine to create important conditions on Earth. If small changes occur in the rate at which water is cycled into or out of a reservoir, the volume of a reservoir changes. These volume changes may be relatively large and rapid in a small reservoir or small and slow in a large reservoir. A small percentage change in the volume of the oceans may produce a large proportional change in the land-ice reservoir, thereby promoting glacial and interglacial stages. The rate at which water enters or leaves a reservoir divided into the reservoir volume determines the residence time of water in the reservoir. The residence time of water in a reservoir, in turn, governs many of the properties of that reservoir.

The huge volume of water contained in the oceans (and seas), $13^7 \times 10^7$ cubic km (about 33×10^7 cubic miles), has been produced during Earth's geologic history. There is little information on the early history of Earth's waters. However, fossils dated from the Precambrian some 3.3 billion years ago show that bacteria and cyanobacteria (blue-green algae) existed then, indicating the presence of water during that period. Carbonate sedimentary rocks, obviously laid down in an aquatic environment, have been dated to 1 billion years ago. Also, there is fossil evidence of primitive marine algae and invertebrates from the Ediacaran Period (635 million to 541 million years ago).

The presence of water on Earth at even earlier times is not documented by physical evidence. It has been suggested, however, that the early hydrosphere formed in response to condensation from the early atmosphere. The ratios of certain chemical elements on Earth indicate that the planet formed by the accumulation of cosmic dust and was slowly warmed by radioactive and compressional heating. This heating led to the gradual separation and migration of materials to form Earth's core, mantle, and crust. The early atmosphere is thought to have been highly reducing and rich in gases, notably in hydrogen, and to include water vapour.

Earth's surface temperature and the partial pressures of the individual gases in the early atmosphere affected the atmosphere's equilibration with the terrestrial surface. As time progressed and the planetary interior continued to warm, the composition of the gases escaping from within Earth gradually changed the properties of its atmosphere, producing a gaseous mixture rich in carbon dioxide (CO_2), carbon monoxide (CO), and molecular nitrogen (N_2). Photodissociation (i.e., separation due to the energy of light) of water vapour into molecular hydrogen (H_2) and molecular oxygen (O_2) in the upper atmosphere allowed the hydrogen to escape and led to a progressive increase of the partial pressure of oxygen at Earth's surface. The reaction of this oxygen with the materials of the surface gradually caused the vapour pressure of water vapour to increase to a level at which liquid water could form. This water in liquid form accumulated in isolated depressions of Earth's surface, forming the nascent oceans. The high carbon dioxide content of the atmosphere at this time would have allowed a buildup of dissolved carbon dioxide in the water and made these early oceans acidic and capable of dissolving surface rocks that would add to the water's salt content. Water must have evaporated and condensed rapidly and accumulated slowly at first. The required buildup of atmospheric oxygen was slow because much of this gas was used to oxidize methane, ammonia, and exposed rocks high in iron. Gradually, the partial pressure of the oxygen gas in the atmosphere rose as photosynthesis by bacteria and photodissociation continued to supply oxygen. Biological processes involving algae increased, and they gradually decreased the carbon dioxide content and increased the oxygen content of the atmosphere until the oxygen produced by biological processes outweighed that produced by photodissociation. This, in turn, accelerated the formation of surface water and the development of the oceans.

Relative Distribution of the Oceans

Earth possesses one "world ocean." However, those conducting oceanic research generally recognize the existence of five major oceans: the Pacific, Atlantic, Indian, Arctic, and Southern oceans. Arbitrary boundaries separate these bodies of water. The boundaries of each ocean are largely defined by the continents that frame them. In the Southern Hemisphere the southern portions of the Pacific, Atlantic, and Indian oceans and their tributary seas that surround Antarctica are often referred to as the Southern Ocean. Many subdivisions can be made to distinguish the limits of seas and gulfs that have historical, political, and sometimes ecological significance. However, water properties, ocean currents, and biological populations are not constrained by these boundaries. Indeed, many researchers do not recognize them either.

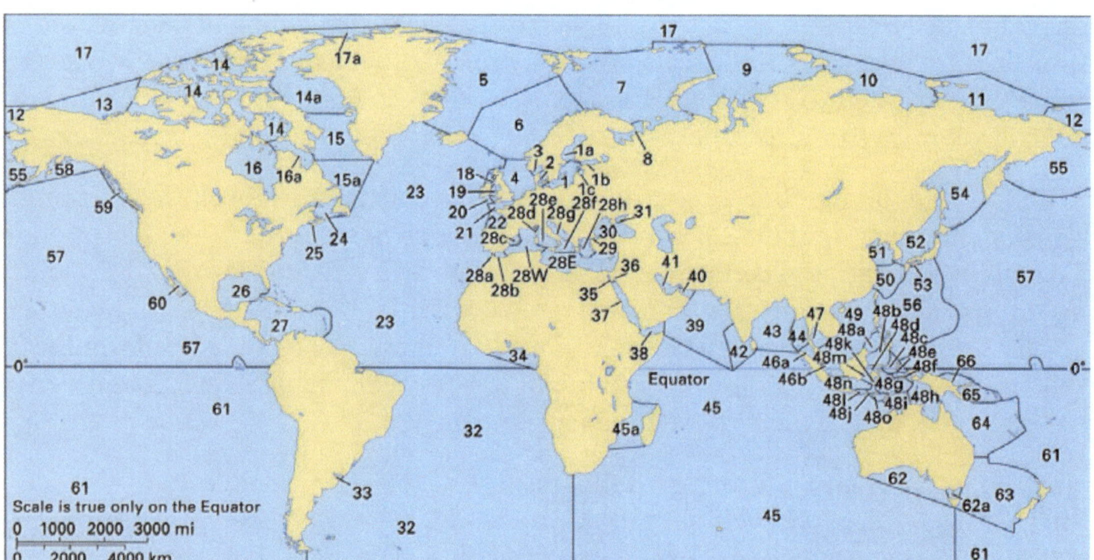

Oceans and seas: boundaries Boundaries of the world's oceans and seas.

If area-volume analyses of the oceans are to be made, then boundaries must be established to separate individual regions. This work was updated in 1966 by American geologist H.W. Menard and American oceanographer S.M. Smith. The latter only slightly changed the numbers derived by Kossina. This was remarkable, since the original effort relied entirely on the sparse depth measurements accumulated by individual wire soundings, while the more recent work had the benefit of acoustic depth soundings collected since the 1920s. This type of analysis, called hypsometry, allows quantification of the surface area distribution of the oceans and their marginal seas with depth.

The distribution of oceanic surface area with 5° increments of latitude shows that the distribution of land and water on Earth's surface is markedly different in the Northern and Southern hemispheres. The Southern Hemisphere may be called the water hemisphere, while the Northern Hemisphere is the land hemisphere. This is especially true in the temperate latitudes.

This asymmetry of land and water distribution between the Northern and Southern hemispheres makes the two hemispheres behave very differently in response to the annual variation in solar radiation received by Earth. The Southern Hemisphere shows only a small change in surface temperature from summer to winter at temperate latitudes. This variation is controlled primarily by the ocean's response to seasonal changes in heating and cooling. The Northern Hemisphere has one change in surface temperature controlled by its oceanic area and another controlled by its land area. In the temperate latitudes of the Northern Hemisphere, the land is much warmer than the oceanic area in summer and much colder in winter. This situation creates large-scale seasonal changes in atmospheric circulation and climate in the Northern Hemisphere that are not found in the Southern Hemisphere.

Major Subdivisions of the Oceans

If the volume of an ocean is divided by its surface area, the mean depth is obtained. Even without including its marginal seas, the Pacific is the largest ocean in both surface area and volume, the Atlantic is next, and the Arctic is the smallest. The Atlantic exhibits the largest change in surface area and volume when its marginal seas are subtracted. This indicates that the Atlantic has the greatest area of bordering seas, many of which are shallow.

Surface area, volume, and average depth of oceans						
Ocean	Area		Volume		Average depth	
	000,000 sq km	000,000 sq mi	000,000 cu km	000,000 cu mi	m	ft
Atlantic Ocean	85.133	32.870	310.411	74.472	3,646	11,962
Pacific Ocean	168.723	65.144	669.880	160.713	3,970	13,025
Indian Ocean	70.560	27.243	264.000	63.337	3,741	12,274
Arctic Ocean	15.558	6.007	18.750	4.498	1,205	3,953
Southern Ocean	21.960	8.479	71.800	17.226	3,270	10,728
the world ocean	361.934	139.743	1,335.000	320.284	3,688	12,100

Hypsometry can show how the area of each ocean or marginal sea changes as depth changes. A special curve known as a hypsometric, or hypsographic, curve can be drawn that portrays how the surface area of Earth is distributed with elevation and depth. This curve has been drawn to represent the total Earth and all of its oceans; likewise, curves can be constructed for each individual ocean and sea. The average depth of the world's oceans, 3,688 metres (12,100 feet), and the average elevation of the land, 840 metres (2,756 feet), are indicated. The highest point on land, Mount Everest (8,850 metres [29,035 feet]), and the deepest point in the ocean, located in the Mariana Trench (11,034 metres [36,201 feet]), mark the upper and lower limits of the curve, respectively. Since this curve is drawn on a grid of elevation versus Earth's area, the area under the curve covering the 29.2 percent of Earth's surface that is above sea level is the volume of land above sea level. Similarly, the area between sea level and the curve depicting the remaining 70.8 percent of Earth's surface below sea level represents the volume of water contained in the oceans.

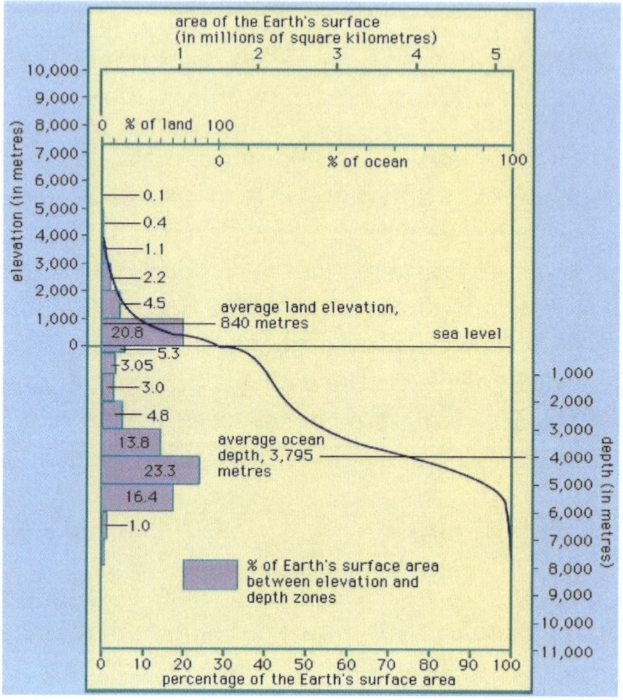

Hypsographic curve showing how the surface area of Earth is distributed with elevation and depth.

Portions of this curve describe the area of Earth's surface that exists between elevation or depth increments. On land, little of Earth's total area—only about 4 percent—is at elevations above 2,000 metres (about 6,560 feet). Most of the land, 25.3 percent of the total Earth, is between 0 and 2,000 metres. About 13.6 percent of the total land area is at higher elevations, with 86.4 percent between 0 and 2,000 metres when the areas are determined relative to land area only. In the oceans the percentages of the area devoted to depth increments yield information about the typical structure and shape of the oceanic basins. The small depth increment of 0–200 metres (656 feet) occupies about 5.4 percent of Earth's total area or 7.6 percent of the oceans' area. This approximates the world's area of continental shelves, the shallow flat borderlands of the continents that have been alternately covered by the oceans during interglacial stages and uncovered during glacial periods.

At depths between 200 and 2,000 metres, an area only slightly larger—6.02 percent of Earth's total area or 8.5 percent of the oceans' area—is found. These depths are related to the regions of the oceans that have very steep slopes where depth increases rapidly. These are the continental slope regions that mark the true edge of the continental landmasses. Marginal seas of moderate depths and the tops of seamounts, however, add their area to these depth zones when all the oceans are considered. The majority of the oceanic area lies between 4,000 and 5,000 metres (about 13,100 and 16,400 feet).

The continental shelf region varies immensely from place to place. The seaward boundary of the continental shelf historically is determined by the 100-fathom, or 200-metre, depth contour. However, 85 fathoms, or 170 metres [about 560 feet], is a closer approximation. The true boundary at any given location is marked by a rapid change in slope of the seafloor known as the shelf break. This change in slope may be nearly at the coastline in areas where crustal plates converge, as along the west coast of North and South America, or it may be located more than 1,000 km (about 620 miles) seaward of the coast, as off the north coast of Siberia. The average width of the shelf is about 75 km (about 45 miles), and the shelf has an average slope of about 0.01°, a slope that is barely discernible to the human eye. Seaward of the shelf break, the continental slope is inclined by about 4°.

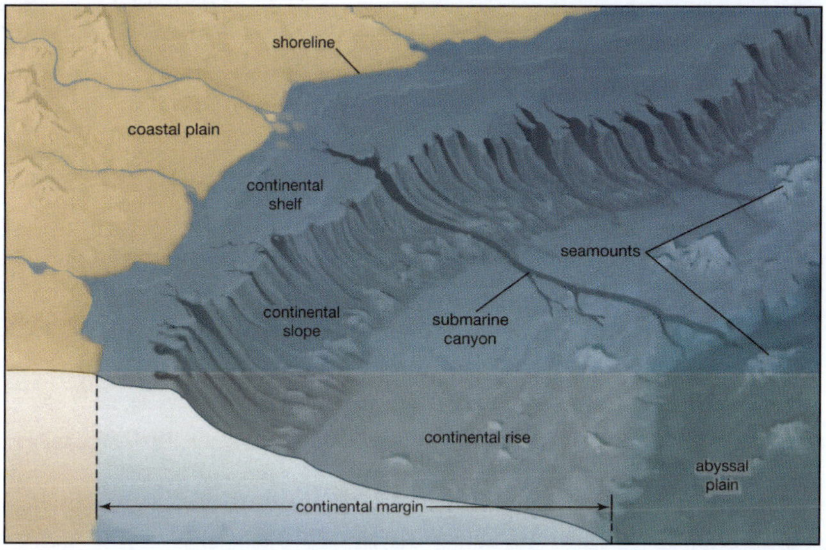

Continental margin: The broad, gentle pitch of the continental shelf gives way to the relatively steep continental slope. The more gradual transition to the abyssal plain is a sediment-filled region called the continental ise. The continental shelf, slope, and rise are collectively called the continental margin.

Intertidal Zone

The intertidal zone is the area of the marine shoreline that is exposed to air at low tide, and covered with seawater when the tide is high. Intertidal zonation refers to the tendency of plants and animals to form distinct communities between the high and low tide lines. All the plants and animals that inhabit this zone are marine, but they have adapted to survive in this often challenging environment. At low tide they must be able to survive prolonged exposure to the air, large fluctuations in temperature and salinity, extremes in wave action, as well as predation from land animals (e.g. raccoons, birds) during low tide, and marine predators like sea stars and fish when the tide is in. Starvation is also a risk when the tide is out since most intertidal animals feed only when they are submerged. Also, most intertidal animals need to be underwater to respire and this can be a limiting factor affecting their distribution on the shore. These challenges are more pronounced higher on the shore where organisms are exposed for longer periods of time. On the positive side, animals such as mussels and barnacles can escape predation from sea stars by inhabiting the upper intertidal zone, above the limit where sea stars can thrive.

Types of species of Flora and Fauna found in Intertidal Zone

The marine waters of the Pacific Northwest support a rich and diverse intertidal marine community, due in part to upwelling currents that bring nutrient-rich waters up to the surface. Despite the environmental challenges of the intertidal zone, there is a wide diversity of marine life forms that inhabit this ecosystem. Some types of intertidal species and their adaptation strategies include:

- Mobile species such as crabs, limpets and snails hide under rocks, in moist crevices or under seaweeds to escape the sun and keep from drying out.
- Some attached animals like barnacles and mussels can hold a small amount of seawater within their shells, and close up tightly during low tide.
- Algae such as Porphyra, Fucus and Enteromorpha have the remarkable ability to survive losing 60-90% of their moisture, to the point of becoming brittle; when the tide comes back in, they reabsorb water.
- Many non-motile (sedentary) species are adapted to withstand high wave action. For example, kelps attach to the substrate with strong holdfasts, barnacles stick to rocks by secreting cement, and sea stars have thousands of tube feet that hug the shore with tiny suction cups. Mussels secrete strong pliable "byssus" threads that anchor them to the shore.

Plants and animals with similar tolerances to the stresses of intertidal life tend to form communities or bio-bands along the shoreline. These horizontal bands are often easy to spot even from a distance, particularly on rocky shorelines, due to differences in colour and form. For example, just below the trees there is typically a black-grey band of encrusting algae and lichens; slightly below that will be a greyish band of barnacles with a yellowish-brown band of rockweed (Fucus) and mussels below that.

Zones along marine shores are generally divided into the following. The backshore is the upper area of land between the high tide line and an imaginary boundary 50 meters inland. Terrestrial species like forests and land animals dominate this zone.

Closest to the high tide line, the backshore zone is fringed with the spray zone. This area is not covered by the high tide, but it is exposed to salt spray. It is typically inhabited by periwinkle snails (Littorina), black encrusting lichens (Verrucaria) and orange lichen (Caloplaca).

The intertidal zone is the broadest zone. It lies between the highest high tide and the lowest low tide, and is flooded once or twice daily by the tide. The form and shape of a shoreline, as well as the degree of wave exposure determine what species of animals and plants are likely to be found living there. Rocky shores tend to be steeper than sediment shores, and some fine sediment shores, such as Patricia Bay on Saanich Inlet, have very wide shallow-sloped intertidal areas.

On a rocky shore, the intertidal zone is marked at its upper limit by barnacles, and closest to the low tide by large kelps (e.g. Laminaria and bull kelp). Other species present include Fucus (rockweed), barnacles, limpets, Nucella (a snail), hermit crabs, shore crabs, mussels, anemones, chitons, sea stars and many seaweed species, to name a few. On shores with fine sediments (sand, mud, small pebbles), many animals are found buried in the sediment or attached to pebbles and cobbles. These include clams, mud shrimp, sand dollars and many types of worms.

The subtidal fringe is exposed only by the lowest of low tides. It borders the subtidal zone, which is always submerged. The subtidal fringe has the richest diversity of species of all the intertidal zones, and includes those that can tolerate only a short exposure to air. Some of the many types of species in this zone include kelp, eelgrass, many species of red algae, sea slugs, sponges, bryozoans, snails, limpets, chitons, crabs, shrimp, sea anemones, sea stars, sea urchins, sea cucumbers, moon snails as well as fish such as sculpins, clingfish and gunnels.

Importance of Intertidal Zone

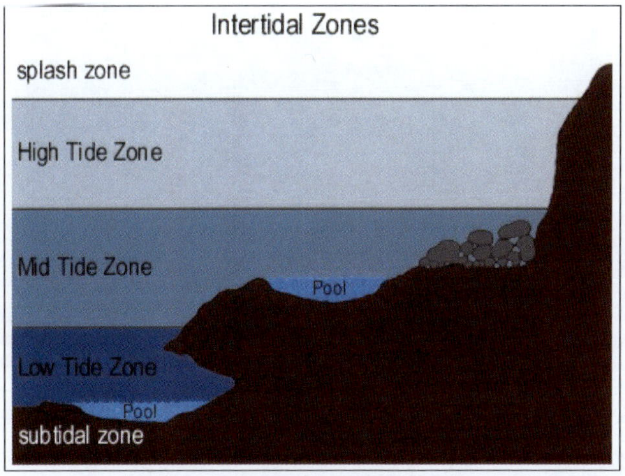

Intertidal zones.

The intertidal zone is a critical interface between terrestrial and marine ecosystems. In this region, characterized by mixed semi-diurnal tides (two high and low tides in 24hr, of unequal size), the ebb and flow of the tide determines the rhythm of life cycles and behaviors of many animals, both marine and terrestrial.

- For migrating birds, low tides periods are critical feeding times when birds have access to the rich food that mudflats, estuaries and similar intertidal habitats provide.

- Resident birds such as herons, eagles, gulls, and many species of shorebird feed in the intertidal zone during low tides.
- Animals such as bears, wolves, raccoons, mink, otter and deer forage for a variety of marine life during low tide.
- For thousands of years, intertidal animals and plants have provided an important food source for many human cultures. In many parts of the BC coast, aquaculture of species such as the Pacific oyster depends on a clean and healthy intertidal zone.
- The intertidal zone provides a physical buffer protecting land from erosion by wave action during storms. This protection is critical for buildings, roads, bridges and other infrastructure built along shorelines.
- A rocky shoreline at low tide provides an irreplaceable outdoor classroom for students of all ages. In this region of the Pacific northwest, we are fortunate to have a rich and diverse intertidal zone, and conservation and environmental stewardship of this ecosystem is essential.

Coral Reefs

Animal, Vegetable and Mineral

The brownish-green specks are the zooxanthellae that most shallow, warm-water corals depend on for much of their food.

Corals are related to sea anemones, and they all share the same simple structure, the polyp. The polyp is like a tin can open at just one end: the open end has a mouth surrounded by a ring of tentacles. The tentacles have stinging cells, called nematocysts, that allow the coral polyp to capture small organisms that swim too close. Inside the body of the polyp are digestive and reproductive tissues. Corals differ from sea anemones in their production of a mineral skeleton.

Shallow water corals that live in warm water often have another source of food, the zooxanthellae (pronounced zo-o-zan-THELL-ee). These single-celled algae photosynthesize and pass some of the food they make from the sun's energy to their hosts, and in exchange the coral animal gives nutrients to the algae. It is this relationship that allows shallow water corals to grow fast enough to build the enormous structures we call reefs. The zooxanthellae also provide much of the green, brown, and reddish colors that corals have. The less common purple, blue, mauve colors found in some corals the coral makes itself.

Coral Diversity

Flower-like clusters of pink polyps make up this coral colony.

In the so-called true stony corals, which compose most tropical reefs, each polyp sits in a cup made of calcium carbonate. Stony corals are the most important reef builders, but organpipe corals, precious red corals, and blue corals also have stony skeletons. There are also corals that use more flexible materials or tiny stiff rods to build their skeletons—the seafans and sea rods, the rubbery soft corals, and the black corals.

The family tree of the animals we call corals is complicated, and some groups are more closely related to each other than are others. All but the fire corals (named for their strong sting) are anthozoans, which are divided into two main groups. The hexacorals (including the true stony corals and black corals, as well as the sea anemones) have smooth tentacles, often in multiples of six, and the octocorals (soft corals, seafans, organpipe corals and blue corals) have eight tentacles, each of which has tiny branches running along the sides. All corals are in the phylum Cnidaria, the same as jellyfish.

Reproduction

A purple hard coral releases bundles of pink eggs glued together with sperm.

Corals have multiple reproductive strategies – they can be male or female or both, and can reproduce either asexually or sexually. Asexual reproduction is important for increasing the size of the colony, and sexual reproduction increases genetic diversity and starts new colonies that can be far from the parents.

Asexual Reproduction

Asexual reproduction results in polyps or colonies that are clones of each other - this can occur through either budding or fragmentation. Budding is when a coral polyp reaches a certain size and divides, producing a genetically identical new polyp. Corals do this throughout their lifetime. Sometimes a part of a colony breaks off and forms a new colony. This is called fragmentation, which can occur as a result of a disturbance such as a storm or being hit by fishing equipment.

Sexual Reproduction

In sexual reproduction, eggs are fertilized by sperm, usually from another colony, and develop into a free-swimming larva. There are two types of sexual reproduction in corals, external and internal. Depending on the species and type of fertilization, the larvae settle on a suitable substrate and become polyps after a few days or weeks, although some can settle within a few hours.

Most stony corals are broadcast spawners and fertilization occurs outside the body (external fertilization). Colonies release huge numbers of eggs and sperm that are often glued into bundles (one bundle per polyp) that float towards the surface. Spawning often occurs just once a year and in some places is synchronized for all individuals of the same species in an area. This type of mass spawning usually occurs at night and is quite a spectacle. Some corals brood their eggs in the body of the polyp and release sperm into the water. As the sperm sink, polyps containing eggs take them in and fertilization occurs inside the body (internal fertilization). Brooders often reproduce several times a year on a lunar cycle.

Coral Growth

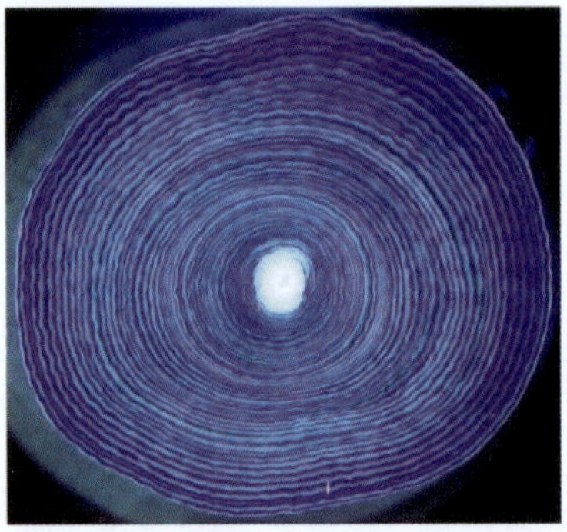

Ultraviolet light illuminates growth rings in a cross-section of 44-year-old Primnoa resedaeformis coral found about 400 m (1,312 ft) deep off the coast of Newfoundland.

Individual coral polyps within a reef are typically very small—usually less than half an inch (or ~1.5 cm) in diameter. The largest polyps are found in mushroom corals, which can be more than 5 inches across. But because corals are colonial, the size of a colony can be much larger: big mounds can be the size of a small car, and a single branching colony can cover an entire reef.

Reefs, which are usually made up of many colonies, are much bigger still. The largest coral reef is the Great Barrier Reef, which spans 1,600 miles (2,600 km) off the east coast of Australia. It is so large that it can be seen from space.

Reefs form when corals grow in shallow water close to the shore of continents or smaller islands. The majority of coral reefs are called fringe reefs because they fringe the coastline of a nearby landmass. But when a coral reef grows around a volcanic island something interesting occurs. Over millions of years, the volcano gradually sinks, as the corals continue to grow, both upward towards the surface and out towards the open ocean. Over time, a lagoon forms between the corals and the sinking island and a barrier reef forms around the lagoon. Eventually, the volcano is completely submerged and only the ring of corals remains. This is called an atoll. Waves may eventually pile sand and coral debris on top of the growing corals in the atoll, creating a strip of land. Many of the Marshall Islands, a system of islands in the Pacific Ocean and home to the Marshallese, are atolls.

It takes a long time to grow a big coral colony or a coral reef, because each coral grows slowly. The fastest corals expand at more than 6 inches (15 cm) per year, but most grow less than an inch per year. Reefs themselves grow even more slowly because after the corals die, they break into smaller pieces and become compacted. Individual colonies can often live decades to centuries, and some deep-sea colonies have lived more than 4000 years. One way we know this is because corals lay down annual rings, just as trees do. These skeletons can tell us about what conditions were like hundreds or thousands of years ago. The Great Barrier Reef as it exists today began growing about 20,000 years ago.

Where are Reefs Found?

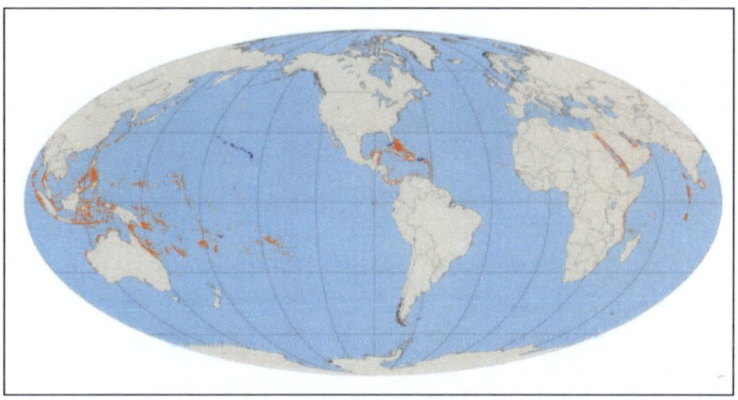

Corals are found across the world's ocean, in both shallow and deep water, but reef-building corals are only found in shallow tropical and subtropical waters. This is because the algae found in their tissues need light for photosynthesis and they prefer water temperatures between 70-85 °F (22-29 °C).

There are also deep-sea corals that thrive in cold, dark water at depths of up to 20,000 feet (6,000 m). Both stony corals and soft corals can be found in the deep sea. Deep-sea corals do not have the same algae and do not need sunlight or warm water to survive, but they also grow very slowly. One place to find them is on underwater peaks called seamounts.

Reefs as Ecosystems

Cities of the Sea

Reefs are the big cities of the sea. They exist because the growth of corals matches or exceeds the death of corals – think of it as a race between the construction cranes (new coral skeleton) and the wrecking balls (the organisms that kill coral and chew their skeletons into sand).

When corals are babies floating in the plankton, they can be eaten by many animals. They are less tasty once they settle down and secrete a skeleton, but some fish, worms, snails and sea stars prey on adult corals. Crown-of-thorns sea stars are particularly voracious predators in many parts of the Pacific Ocean. Population explosions of these predators can result in a reef being covered with tens of thousands of these starfish, with most of the coral killed in less than a year.

Scientists have been studying why populations of crown-of-thorns sea stars (Acanthaster planci) have mushroomed in recent decades.

Corals also have to worry about competitors. They use the same nematocysts that catch their food to sting other encroaching corals and keep them at bay. Seaweeds are a particularly dangerous competitor, as they typically grow much faster than corals and may contain nasty chemicals that injure the coral as well.

Corals do not have to only rely on themselves for their defenses because mutualisms (beneficial relationships) abound on coral reefs. The partnership between corals and their zooxanthellae is one of many examples of symbiosis, where different species live together and help each other. Some coral colonies have crabs and shrimps that live within their branches and defend their home against coral predators with their pincers. Parrotfish, in their quest to find seaweed, will often bite off chunks of coral and will later poop out the digested remains as sand. One kind of goby chews up a particularly nasty seaweed, and even benefits by becoming more poisonous itself.

Threats

Global

These bleached corals in the Gulf of Mexico are the result of increased water temperatures.

The greatest threats to reefs are rising water temperatures and ocean acidification linked to rising carbon dioxide levels. High water temperatures cause corals to lose the microscopic algae that produce the food corals need—a condition known as coral bleaching. Severe or prolonged bleaching can kill coral colonies or leave them vulnerable to other threats. Meanwhile, ocean acidification means more acidic seawater, which makes it more difficult for corals to build their calcium carbonate skeletons. And if acidification gets severe enough, it could even break apart the existing skeletons that already provide the structure for reefs. Scientists predict that by 2085 ocean conditions will be acidic enough for corals around the globe to begin to dissolve. For one reef in Hawaii this is already a reality.

Local

Lionfish are referred to as turkeyfish because, depending on how you view them, their spines can resemble the plumage of a turkey.

Unfortunately, warming and more acid seas are not the only threats to coral reefs. Overfishing and overharvesting of corals also disrupt reef ecosystems. If care is not taken, boat anchors and divers can scar reefs. Invasive species can also threaten coral reefs. The lionfish, native to Indo-Pacific waters, has a fast-growing population in waters of the Atlantic Ocean. With such large numbers the fish could greatly impact coral reef ecosystems through consumption of, and competition with, native coral reef animals.

Even activities that take place far from reefs can have an impact. Runoff from lawns, sewage, cities, and farms feeds algae that can overwhelm reefs. Deforestation hastens soil erosion, which clouds water—smothering corals.

Coral Bleaching

Compare the healthy coral on the left with the bleached coral on the right.

"Coral bleaching" occurs when coral polyps lose their symbiotic algae, the zooxanthellae. Without their zooxanthellae, the living tissues are nearly transparent, and you can see right through to the stony skeleton, which is white, hence the name coral bleaching. Many different kinds of stressors can cause coral bleaching – water that is too cold or too hot, too much or too little light, or the dilution of seawater by lots of fresh water can all cause coral bleaching. The biggest cause of bleaching today has been rising temperatures caused by global warming. Temperatures more than 2 degrees F (or 1 degree C) above the normal seasonal maximimum can cause bleaching. Bleached corals do not die right away, but if temperatures are very hot or are too warm for a long time, corals either die from starvation or disease. In 1998, 80 percent of the corals in the Indian Ocean bleached and 20 percent died.

Protecting Coral Reefs

A bluefin trevally swims in Hawaii's Maro.

There is much that we can do locally to protect coral reefs, by making sure there is a healthy fish community and that the water surrounding the reefs is clean. Well-protected reefs today typically have much healthier coral populations, and are more resilient (better able to recover from natural disasters such as typhoons and hurricanes).

Fish play important roles on coral reefs, particularly the fish that eat seaweeds and keep them from smothering corals, which grow more slowly than the seaweeds. Fish also eat the predators of corals, such as crown of thorns starfish. Marine protected areas (MPAs) are an important tool for keeping reefs healthy. Large MPAs protect the Great Barrier Reef and the Northwestern Hawaiian Islands, for example, and in June 2012, Australia created the largest marine reserve network in the world. Smaller ones, managed by local communities, have been very successful in developing countries.

Clean water is also important. Erosion on land causes rivers to dump mud on reefs, smothering and killing corals. Seawater with too many nutrients speeds up the growth of seaweeds and increases the food for predators of corals when they are developing as larvae in the plankton. Clean water depends on careful use of the land, avoiding too many fertilizers and erosion caused by deforestation and certain construction practices. In the long run, however, the future of coral reefs will depend on reducing carbon dioxide in the atmosphere, which is increasing rapidly due to burning of fossil fuels. Carbon dioxide is both warming the ocean, resulting in coral bleaching, and changing the chemistry of the ocean, causing ocean acidification. Both making it harder for corals to build their skeletons.

Lagoon

A lagoon is an area of relatively shallow, quiet water situated in a coastal environment and having access to the sea but separated from the open marine conditions by a barrier. The barrier may be either a sandy or shingly wave-built feature (such as a sandbar or a barrier island), or it may be a coral reef. Thus, there are two main types of lagoons: (1) elongated or irregular stretches of water that lie between coastal barrier islands and the shoreline and (2) circular or irregular stretches of water surrounded by coral atoll reefs or protected by barrier coral reefs from direct wave action. Lagoon depths are maintained at a moderate level by sedimentation, and this compensates for the subsidence that commonly attends reef formation. Because the reef is an organic structure, the lagoonal sediments contain much calcareous material. The sheltered waters support highly productive ecosystems made up of a distinctive flora and fauna.

Lagoon Types

Barrier Island Lagoons

Barrier island, or coastal, lagoons are characterized by quiet water conditions, fine-grained sedimentation, and, in many cases, brackish marshes. Water movements are related to discharge of river flow through the lagoon and to the regular influx and egress of tidal waters through the inlets that normally separate the barrier islands. Coastal lagoons are generally characteristic of coasts of low or moderate energy, occurring especially on the east coasts of continents where the swells are less violent and in high latitudes where offshore ice provides some protection. They also are associated with low coasts and rarely occur where high cliffs form the coast. They can form only where there is abundant sediment for construction of the protective barrier islands. Too much sediment from the mainland, however, can lead to delta formation rather than lagoons, although lagoons frequently occur along the outer delta margin and between delta distributaries.

Gippsland Lakes:
Gippsland Lakes (lagoons), Victoria, Australia.

Coastal lagoons are widely distributed throughout the world and have been estimated to constitute about 13 percent of the total world coastline. Lagoons are more common on coasts with moderate to low tidal ranges; for example, they occur widely on low coasts of the southern Baltic, the southeast North Sea, the Black Sea, the Caspian Sea, and the Mediterranean Sea, as well as on low coasts of the southeastern United States and the Gulf of Mexico. Lagoon coasts also occur along southern Brazil, the east coast of Madagascar, northeastern Russia, Japan, and isolated parts of Africa, India, Australia, and New Zealand.

Coral Lagoons

Coral lagoons are restricted to tropical open seas that provide the conditions necessary for coral growth. They are best exemplified by the roughly circular quiet waters that are surrounded by warm-water coral atoll reefs. Coral lagoons occur widely in the western Pacific, in parts of the Indian Ocean, and in isolated places in the Caribbean, mainly within 25° latitude of the Equator. Coral lagoons are of great importance to many island communities in the Pacific, particularly where they provide the only quiet water for use as harbours, although the passage through the reef into the lagoon is often perilous.

Tuamotu Archipelago: Fakarava island
Lagoon near Rotoava, Fakarava island, Tuamotu Archipelago, French Polynesia.

Coral-reef lagoons also occur on marginal reefs such as the Great Barrier Reef of Australia, but the most spectacular examples are the atolls of the Pacific Ocean, some of which are more than 50 km (30 miles) across. Some atolls consist only of a lagoon, often with a fairly uniform depth, surrounded by a low-lying coral reef; some include one or more high, rocky volcanic islands, and others are complex, with small reefs surrounded by lagoons within a larger reef. All are thought to have been built by the upward growth of coral during a relative rise in sea level due to subsidence and eustatic (global) change.

Nature of the Lagoon Environment

Dimensions

Coral lagoon dimensions range from small atolls to those so wide that the coral reefs on the far side cannot be seen across the lagoon. Atoll widths range from about 2.5 to nearly 100 km (1.5 to nearly 62 miles), but the mean value is about 20 km (about 12 miles). Depths rarely exceed 60 metres (about 200 feet) and many are shallower, usually less than 20 metres (about 65 feet) deep. The lagoon of Mayotte island in the Comoro archipelago in the Indian Ocean attains a maximum depth of about 92 metres (about 300 feet), but it is generally shallower. That lagoon is about 16 km (about 10 miles) in width at its widest point and lies behind a barrier reef that encircles the island, forming an atoll about 55 km (about 34 miles) in diameter.

Barrier island lagoons are usually elongated, though irregular ones may occur where river estuaries flood behind barriers. This occurs on the east coast of the United States, where lagoons extend intermittently for nearly 1,500 km (about 900 miles) along the coast. The Gippsland Lakes in Victoria, Australia, exemplify a complex lagoon system formed behind a 149-km (93-mile) beach. Elongated lagoons up to 64 km (about 40 miles) in length lie behind the beach barrier, and larger lagoons, such as Lake Wellington, lie behind the southwestern end. Postglacial subsidence has flooded the lowland in this area. The lagoons are shallow: Lake Wellington is less than 3.5 metres (11.5 feet) in depth, and much of Lake King is less than 6 metres (about 20 feet) deep. Scour holes as deep as 16.5 metres (about 54 feet) do occur, however. The elongated lagoon behind the barrier is only 1 to 1.5 metres (3 to 5 feet) deep, typical of barrier island lagoons.

Water Circulation

The degree of water circulation depends on the width of the tidal inlets, the tidal range, and the amount of runoff from adjacent land areas. Maximum velocities are attained at the points where the water passes through the barriers. In the entrance to the Gippsland lagoons, for example, tidal currents reach 5.6 km (3.5 miles) per hour, but river floods that escape to the ocean can raise the velocity to 13 km (8 miles) per hour. Water may be blown into the lagoon by strong winds; the increased level results in an outflowing current when the wind drops. Seiches—that is, rhythmic oscillations of water in enclosed or partially enclosed water bodies—can be created in this way. Small waves can be generated within lagoons when the wind blows along their maximum dimension. These may reach 1.25 metres (about 4 feet) in height and 1.5 to 9 metres (about 5 to 30 feet) in length in the Gippsland lagoons. In coral atoll lagoons there is little or no runoff, and seawater moves in and out through the passes, where tidal currents reach their maxima. Velocities of 19 to 22 km (12 to 14 miles) per hour have been recorded in the Hao Channel of the Tuamotu Archipelago in French Polynesia.

A lagoon into which a major river flows is known as an estuarine lagoon and may be regarded as a special kind of estuary.

Water Temperature and Salinity

In the Mayotte lagoon the water has the same properties as the upper layers of the open ocean. The salinity is close to 35 parts per thousand (ppt), and the temperature varies between 27 and 24 °C (81 and 75 °F). This is typical of coral lagoons, but the temperature and salinity of barrier island lagoons are more variable because of their wider climatic range. Because they are shallow, lagoon waters approximate the air temperature: colder than the open ocean in winter, warmer in summer.

Salinities decrease as a function of the amount of runoff entering the lagoon in relation to the tidal influx. Lagoons may be considered brackish, marine, or hypersaline. Brackish lagoons receive much runoff, and salinity increases toward the tidal inlets. The Gippsland lagoons exemplify this type. The salinity at the inner end varies from 0.5 to 5 ppt according to season, and central values vary between 5 and 20 ppt.

In hot arid regions lagoons lose more water by evaporation than they receive from land drainage. This causes surface waters to become more dense than seawater and to sink to the bottom. Seawater flows in at the surface to replace that lost by evaporation, creating a circulation the reverse of that found in estuaries. If exchange with the open sea is limited, the lagoon may become much more saline than the open sea. Consequently, various species of plants and animals become adapted to life in high salinities. Laguna Madre in Texas and Syvash Sound in the Black Sea are examples of hypersaline lagoons. (They have salinities of 65 and 132 ppt, respectively.) Salt deposits may form in these conditions. The denser saline water tends to lie beneath the fresher water, where mixing is not severe.

Equilibrium Bottom Profiles

Lagoons behind coastal barriers normally are zones of fine sedimentation. Their bottom profiles, therefore, build up gradually with time. Typical depths of the Texas lagoons vary between 1.25 and 3.5 metres (4 and 11 feet), and their floors are flat. Early theories that attempted to relate the form of the offshore and lagoon profile are no longer held; and, because the lagoon profile changes with continued deposition, it cannot be used to establish the process of lagoon formation. The profile is usually gently undulating, but it may contain deeper channels, especially near the tidal inlets.

Profiles across coral lagoons either are smooth and flat from calcareous sedimentation or contain knolls of growing or dead coral. There are 2,300 such knobs in the Enewetak Lagoon in the Marshall Islands.

Factors Involved in Lagoon Formation

The essential feature that causes the lagoon to exist is the barrier that separates it from the ocean. In the coral lagoon the formation of the reef depends on the existence of suitable conditions for reef growth.

Barrier Bars and Sediment Sources

The barrier island lagoons, on the other hand, depend not on organic processes but on waves, which act in a suitable environment on an adequate supply of bottom material, most commonly sand. Barrier islands are formed in those areas where the coastal slope is flatter than the equilibrium slope required by the long constructive swells—i.e., the waves that build up the foreshore in front of their breakpoint. They are, therefore, found on low coasts. They may occur in areas of subsidence, stability, or emergence, wherever sufficient sand exists.

The Hotel Zone, which sits on a former barrier island facing the Caribbean Sea, is linked by causeway to the city of Cancún, Mex. The calm blue water of Nichupte Lagoon is on the left.

The barrier islands that form the lagoons are made of sand, but the sediments of the lagoon are usually finer, as conditions are quieter. The lagoonal muds differ from shelf muds. Glauconite is rare in lagoon muds, but oyster reefs may occur as along the Texas coast. The muds are found away from the channels, in which only coarse sediment can rest, owing to strong currents. Flocculation in the saline lagoon water expedites clay deposition. The source of the fine sediment is from inland areas, and transport is by rivers. The details of lagoon sedimentation vary with the nature of the river load. Sedimentation rates are much greater in the lagoon than the adjacent open ocean, because deposition is continuous over much of the lagoon and is often aided by flocculation and vegetation. In the Texas lagoons from 1875 to 1936, 20 cm (about 8 inches) of deposition occurred in spite of 30 cm (about 12 inches) of subsidence; the sedimentation rate, therefore, was about one metre in 100 years.

Waves, Tides and Surf

The barrier islands are formed by the waves, which build up their equilibrium profile by deposition on a gradient that is too flat. The level of the growing accumulation may be raised by the wind, forming dunes. Where the land behind the growing barrier is low, it will become flooded to form a lagoon if sea level rises slowly. Such a rise of sea level has occurred during the past 20,000 years. As long as the barrier island can maintain its level above the sea, as sea level rises, the lagoon will exist until it is filled with sediment. Not all lagoons and barrier island complexes have formed during rising sea-level conditions, but, where there is evidence that no open-sea foreshore sediments lie on the land side of the barrier, this hypothesis seems likely. In some barriers, however, outbuilding of material by glacial outwash streams or rivers may provide a suitably low gradient and enough sediment to form a barrier, as along the south coast of Iceland. In other areas material carried alongshore to form a spit may develop into a bay-mouth barrier, enclosing a lagoon. Such features can be of sand or shingle (pebbles). The Fleet, a brackish body of water behind Chesil Beach in southern England, is an example of the latter type.

Waves within the lagoon may develop cuspate (pointed) spits along the land side of the barrier and the inner edge of the lagoon. These features may eventually break the lagoon into almost circular or oval water bodies. Examples occur in the Chukchi Sea lagoons in northeastern Russia and elsewhere where vegetation does not form marshland.

Storms and Catastrophic Events

Storms and tsunamis exert an effect on lagoons when they breach or overtop the barriers around a lagoon by raising the water level temporarily. Major changes in configuration can occur in a short time. Hurricanes, for instance, can cause serious effects on the coast of Texas, along which lagoons are common. Padre Island was lowered to below mean low tide at its southern end by one storm in 1919, and several washover channels were cut. (Washover channels are inlets of water excavated by the wave action produced by a severe storm.) The mainland shore also suffered erosion.

Deposition may also occur; saline marls have been laid down on freshwater marsh, and small beach ridges may be built inside the lagoon where the high water level drives sand inland over mud. Coral reefs are more resistant to storms than are mobile sandy barriers.

Effect of Time

Lagoons of both types change with time. In both, a relative rise of sea level with time is important in the development of the lagoon. In coral atolls there is evidence from deep boring that English naturalist Charles Darwin's original subsidence hypothesis of atoll formation, via barrier reefs from fringing reefs around a subsiding volcanic peak, is substantially correct in many cases. As long as the coral can maintain its growth at a suitable level as its foundation subsides, the atoll will continue to enclose a lagoon, which is floored by coral or calcareous sediment derived from the reef and which maintains its depth by growth or deposition.

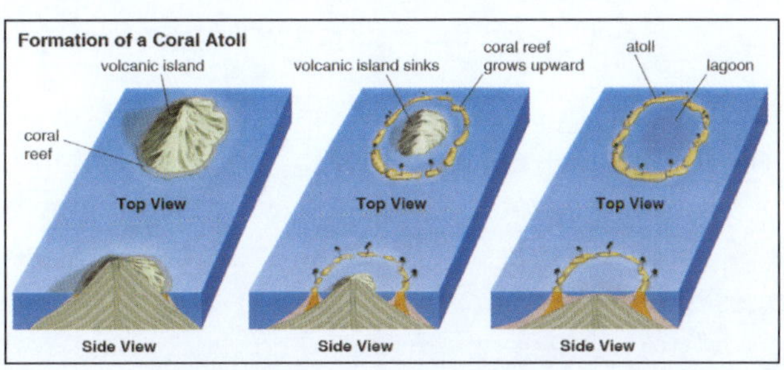

Atoll formation: Formation of an atoll.

The postglacial rise of sea level also has influenced barrier island formation in many instances. When sea level rises too fast, a barrier may be drowned, and its lagoon will cease to exist.

Mangrooves

Mangroves are the only trees that are capable of thriving in salt water. They form unique intertidal forests at the edge of land and sea. They are represented on all continents with tropical and subtropical coasts, i.e. North and South America, Africa and Middle-East, Asia and Oceania.

Mangrove forests or mangals are a type of intertidal wetland ecosystems. The word mangrove is derived from the Portugese word mangue which means “tree” and the English word grove which is used for trees and shrubs that are found in shallow, sandy or muddy areas. They replace Salt marshes in tropical and subtropical regions. They are salt-tolerant forested wetlands at the sea-land interface which forms the link between the terrestrial landscapes and the marine environment. The dominant plants are several species of mangrove (for a species overview, check the Mangrove Species Database). Mangroves are woody trees and shrubs with a thick, partially exposed network of roots that grow down from the branches into the water and sediment. They settle where there is little wave action and where sediments accumulate. While growing, mangal forests further reduce waves and increase sedimentation. The deposits of fine grained (muddy and sandy) sediments lack oxygen. They are frequently associated with saline lagoons and are regularly found on protected sides of islands, atolls and tropical estuaries.

Mangal in Thailand.

The distribution, density and species composition are determined by the water and air temperatures during the winter, exposure to wave action and tidal currents, the range of the tide, the type of sediment and the chemistry of the seawater. The most highly developed and most species rich mangals are found in Malaysia and Indonesia. Over the world, 54-70 species (for a species overview, check the Mangrove Species Database)(and hybrids) in 20-27 genera and 16-19 families are found.

They are almost exclusively tropical, but are also seen in the subtropics. Mangroves are intolerant of frost, but can tolerate air temperatures as low as 5 °C. They are most closely correlated with the seawater temperature. The 20 °C isotherm in the winter is a good indicator for the limit of distribution. The number of species tends to decrease with the distance from the equator. In the southern hemisphere, ranges extend further south on the eastern margins of land masses than on the western. This is because of the pattern of warm and cold ocean currents. But local anomalies of current and temperature or the local evolution can create local changes.

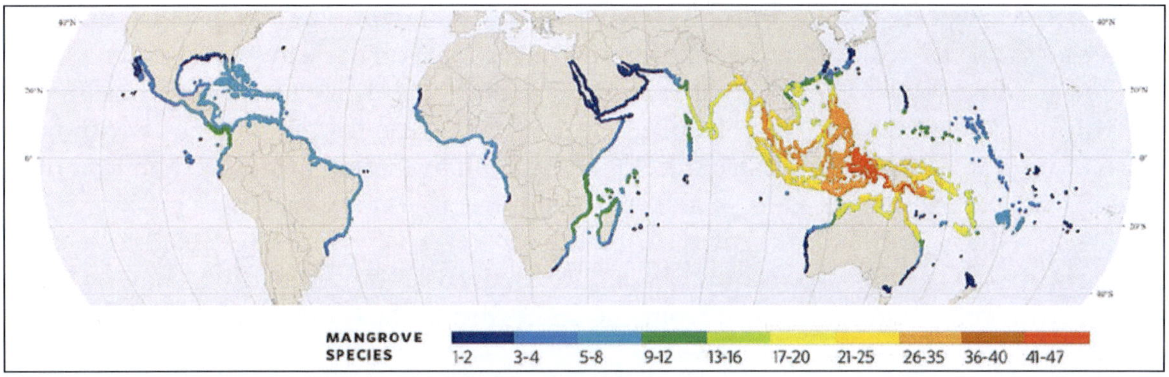

World distribution of mangroves.

Requirements for Development

Mangroves have several requirements to develop:

- They need average temperatures of the coldest month higher than 20 °C. The seasonal temperature range should not exceed 5 °C. They can tolerate temperatures of 5 °C, but the development will be affected. They are not resistant to freezing.
- In general they need a fine-grained substrate. But there could be some exceptions. This is the case in Papua New Guinea and Kenya, where the mangroves grow on corals.
- The shores must be free of strong wave action and tidal current.
- They need salt water. They are facultative halophytes.
- They need a large tidal range. This causes limited erosion and deposition of sediments.

Due to these processes, a well-marked zonation is seen. Each of the zones is dominated by a different mangrove species and associated fauna and flora. Red mangroves (Rhizophora) are usually found the closest to the edge of the water, where the greatest degree of tidal flooding occurs. More landwards are the black mangroves (Avicennia). These areas receive only shallow flooding during high tide. The upper limit of the mangroves is occupied with white mangroves and buttonwoods. The buttonwoods are not really a mangrove species, but are a transition species between the mangrove and the terrestrial vegetation.

Functioning and Adaptations

The mangroves have several functions and adaptations to a life in an intertidal ecosystem. They need to conquer some problems to be resistant to the environment. The first problem is that mangrove trees are freshwater riverine trees. They grow in an environment whose salinity ranges between that of freshwater and seawater. For this reason, they have to take up water against the osmotic pressure. To overcome the negative osmotic pressure, they generate a negative hydrostatic pressure (by transpiration processes). They developed a mechanism to exclude salt by the roots or leaves. In this way, they are tolerant for saline conditions. Even with exclusion of most of the salts, concentration of chloride and sodium ions in the tissue is higher than other plants. This high concentration can inhibit many enzymes. To protect the enzymes, the salt is stored in vacuoles. The high cation concentrations are balanced by high non-ionic solutes in the cytoplasm. Several mangrove species deposit sodium and chloride in the bark of stems and roots. Other species deposit salt in senescent leaves, which later fall off the tree. Salt glands on the leaves also exclude salt. This can be seen as salt crystals. The lower surface of the leave is highly covered with hairs to raise the secreted droplets of salty water away from the leaf surface. This prevents osmotic withdrawal of water from the tissue. They also restrict the opening of their stomata (only on the lower surface of the leaf), have a tick cuticle with a waxy layer, and orientate their leaves to avoid the burning sun. This also reduces evaporation. Because salt tolerance is costly, a greater relative root mass is needed to recover the demand for water. When it rains, drop roots, descending from the branches, absorb the freshwater that runs down from the stem through a special superficial layer. In this way, no high-energy inverse osmosis is needed.

Salt crystals on a mangrove leaf.

Extended root mass.

Drop roots from branches.

A second problem is the anoxic environment. The underground in which they root is saturated with water. The tissue of the plants requires oxygen for respiration. Gas diffusion between gas particles can only supply this need in soils that are not waterlogged. Even when the water is saturated with oxygen, its concentration is too low and the diffusion in water is very slow. This is solved by various forms of aerial roots.

The most roots branch off from the stem underground. One type of roots is the prop root or rhizophore. This root diverges from the tree and anchors into the bottom to stabilize the tree in the soft, muddy substrate. The rhizophore has lenticels on the upper surface. Lenticels are large pores with a corky layer and enable the exchange of air. Seawater cannot get in the lenticels. The tissue of the prop roots consists of aerenchyma and is connected with the lenticels. Through this aerenchyma, air can be provided to the submerged parts of the tree. The root can periodically break the soil surface and submerges again. This forms a knee root.

Knee roots.

Drop roots or rhizophora.

Another type of roots is a shallow, horizontal root that radiates outwards. The vertical root is called a pneumatophore and can be as high as several decimeters. They also have lenticels and aerenchyma. This can create a huge network of vertical roots. The horizontal root is called the cable root.

At last, the buttress root is a root that covers the whole space between the upper part of the root and the bottom.

Pneumatophores on cable roots.

Buttress root.

The pollination of the trees is by the wind or by organisms. All mangroves disperse their offspring by water. They produce unusually large propagating structures or propagules. The embryo initiates germination on the seed, still attached on the tree and further develops into a propagule. This phenomenon is known as vivipary.

Propagules.

Mangroves aid soil formation by trapping debris. Prop roots and pneumatophores accumulate sediments in protected sites and form mangrove peats. The filamentous algae also help to stabilize the fine sediments trapped by mangroves. They usually form a green-to-red mass over the substrate. They are also a filtering system for the land run-off and remove the terrestrial organic matter. They are very important habitats for many species of small fish, invertebrates and various epiflora and epifauna as well as larger birds. This is called a nursery function. The mangrove is a major producer of detritus that will contribute to offshore productivity in some seasons.

Biota

Micro-algae are important in mangrove ecosystems. They are epiphytic and grow on the aerial roots of the trees and on the sediments. The algae are green (Chlorophyta), brown (Phaeophyceae), red (Rhodophyta) and blue-green (Cyanophyta). The dense biomass on the aerial roots causes the water to remain on the pneumatophores. The lenticels are no longer functional and oxygen can not penetrate into the roots. For this reason, the bark regularly falls off the root. This process is called decertification. Vertical zonation along a single pneumatophore occurs, but there is also a zonation from the upper limit of the mangroves to the lower limit.

A number of invertebrates are found in the mangrove ecosystem. Macrobenthic species can be commonly found or they could only occasionally been found because they are migrating within it or because they are living in adjacent environments. Crabs are commonly found in the mangroves. They are keystone species. This means that the presence of this animal in the community makes it possible for many other species to live there. The crabs go through their larval stages in the water beneath the mangroves. When they are mature, they crawl up on the mangroves and feed on the leaves. They are crucial in the processing of the leaf litter. They can reach high densities. Burrow activities also take place. The micro-topography of the bottom will be modified and the soil will be aerated. This decreases the sulphide levels in the soil and positively influences the productivity of the trees. An example of a mangrove crab is the fiddler crab Uca lacteal.

Fiddler crab Uca lacteal.

Egret Egretta alba.

An important bivalve is the purple oyster Lopha frons. This species encrusts the pneumatophores and prop roots. When the tide is high, barnacles and mussels compete with the oyster for space on the roots. Periwinkles also occur on the roots and stems, as well as on the shells of sedentary organisms attached on them. Snails are extremely important in the turn over of the organic material. Other species that occur in the mangroves are tunicates, sponges, ants, hermit crabs, shrimps, fishes. They may be a source of nutrition for higher level predators. Species that cannot tolerate the changing saline conditions can survive in the forest. These species are sea stars, brittle stars and sea squirts. Predators are clapper rails, diamondback turtles, water moccasins, raccoons and killifishes. Bacteria and fungi initially break down the leaf litter (decomposition). In the tree canopy, vertebrate fauna and birds are common. Examples of birds are pelicans, wood ibises, herons, egrets and roseate spoonbills.

Water moccasin Agkistrodon contortrix.

Diamondback turtle Malaclemys terrapin.

Threats

The mangroves are threatened in their existence by several causes. The main source of these threats are induced by humans.

- Variations in river and surface run-off, that inhibit the tropical coastal deltas of fresh water and silt, cause losses of mangrove species diversity and organic production. This results in alternations in both the terrestrial and aquatic food web. This has an effect on the types of refugees available to consumers.
- People will always be engaged in making projects. Soil reclamation for agriculture and aquaculture reduce the regional levels of biodiversity due to loss of mangrove habitats. The shrimp aquaculture for example has major effects on the biodiversity in the mangroves.
- People are clearcutting the mangrove trees (deforestation, habitat loss) and are building dikes. This forms ponds with anoxic water. These anoxic conditions increase the level of sulphide in the soil and increase the pH leading to major shrimp losses.
- Another negative impact of humans on the mangrove habitat is the use of pesticides and fertilizers. The products that are used in the upstream agriculture end up in the water around the mangroves. This causes an increased nutrient concentration, especially nitrogen and phosphorus. These nutrients cause oxygen depletion in the water and promote the growth of algae. As a result, the ecosystems will be no longer in equilibrium.
- Another problem is the clearcutting of the mangrove for their hard wood. This wood is an important export product for building constructions in areas with large concentrations of termites. This is because the wood is resistant against these termites. The wood can also be used as charcoal and fuelwood. The substrate will be no longer stable when the trees are cut away. The result of this unstable condition is erosion.
- A similar effect as with the added fertilizers and pesticides is the use of mangroves in waste-water treatment. Nutrients are added into the water and the equilibrium in the food web is disturbed. Mangroves no longer can survive in this environment and die off. The organic matter, normally stored in the mangroves, will be transported to open water and increases the aquatic primary production. This results in a huge amount of phytoplankton and causes water turbidity. Corals and seagrasses are influenced negatively by these processes and will be deteriorated.
- Other coastal development activities have an influence on the quality of the water. Industry, tourism and port development involve land reclamation and dredging. This causes resuspension of the sediment and makes the water turbid. Because of this, light can not penetrate enough in the water and causes in this way damage to mangroves.
- Spills of oil, toxic chemicals and dumping of waste into the water causes localized impacts on the mangroves. Also the introduction of alien species by ballast water of on the hulls of vessels will have negative effects on the mangrove habitats. They will compete with indigenous species for space and food.

- Another threat is climate change. Storms will become more frequent and more intense and the sea level rises. The storms cause damage to the mangroves and due to the more frequently appearance, the recovery periods will become shorter.

There is little chance of mangrove regeneration in the remaining arid lands.

In the past, several measurements have been taken to preserve the mangrove ecosystem. An example of such a measurement is the FAO Code of Conduct for Responsible Fisheries. The FAO will regulate the fishing techniques and eliminate the destructive fishing gears. They also create management plans, establish no-take areas and involve the fishermen and other users. Mangrove ecosystems can also be protected by setting up National Protected Areas, World Heritage Sites and Ramsar Sites. These special sites have a legal framework. There were created guidelines for the people that use the ecosystem. But there are frequently problems with the implementation of these guidelines. The UNEP Regional Seas Programme is an organization that helps countries to work together to protect the ecosystems. A good management plan is essential to the protection of every ecosystem. In the future, we have to develop much more mechanisms to protect the ecosystem.

Reasons to Protect Mangrove Forests

Mangroves might appear to be merely coastal cousins of inland forests, but these rich ecosystems support the planet and people in unique ways, from providing breeding grounds for fish to carbon storage, to protection against flooding. Yet despite their importance, mangrove forests are under threat. Over a third have already disappeared, and in regions such as the Americas they are being cleared at a faster rate than tropical rainforests.

Much of that clearance is to reclaim land for agriculture, industrial development and infrastructure projects.

In addition to climate change and pollution, there are also local threats. These include overharvesting of wood for fuel and construction, dams and irrigation that reduce the flow of water reaching the forests, and overfishing causing disruption to food chains and fish communities.

We are destroying a coastal ecosystem that helps sustain life and livelihoods.

Here are five of the many reasons we should be doing much more to preserve mangrove forests:

1. They are a natural coastal defence: The sturdy root systems of mangrove trees help form a natural barrier against violent storm surges and floods. River and land sediment is trapped by the roots, which protects coastline areas and slows erosion. This filtering process also prevents harmful sediment reaching coral reefs and seagrass meadows.

In 2017, the UN Ocean Conference estimated that nearly 2.4 billion people live within 100 km of the coast. Mangroves provide valuable protection for communities at risk from sea-level rises and severe weather events caused by climate change.

2. They are carbon sinks: Coastal forests help the fight against global warming by removing carbon dioxide from the atmosphere, most of which is stored within the plant. When mangrove tree roots, branches and leaves die they are usually covered by soil, which is then submerged under tidal water, slowing the breakdown of materials and boosting carbon storage.

Research shows that coastal mangroves outperform most other forests in their capacity to store carbon. An examination of 25 mangrove forests across the Indo-Pacific region found that per hectare, they held up to four times more carbon than other tropical rainforests.

3. They provide livelihoods: Many people living in and around mangroves depend on them for their livelihood. The trees are a reliable source of wood for construction and fuel, which is prized for its hardy resistance to both rot and insects. However, in some areas, the wood has been harvested commercially for pulp, wood chip and charcoal, raising concerns about sustainability.

Plant extracts are collected by locals for their medicinal qualities and the leaves of mangrove trees are often used for animal fodder. The forest waters provide local fishermen with a rich supply of fish, crabs and shellfish to sell for income.

4. They encourage ecotourism: Sustainable tourism offers a stimulus to preserve existing mangrove areas, with potential to generate income for local inhabitants.

Often located near to coral reefs and sandy beaches, the forests provide a rich environment for activities like sports fishing, kayaking and birdwatching tours. Of course, it is important to maintain a balance between visitor numbers and protecting the forests' delicate ecosystem. If held at sustainable levels, ecotourism could provide the perfect motivation to protect mangroves, instead of clearing them for mass tourism developments.

5. They are rich in biodiversity: Human activity has caused huge biodiversity loss in land and marine ecosystems around the globe, endangering many plant and animal species. By filtering coastal waters, mangroves form a nutrient-rich breeding ground for numerous species that thrive above and below the waterline.

A huge variety of wildlife lives or breeds in the mangrove ecosystem, including numerous fish, crab and shrimp species, molluscs, and mammals like sea turtles. The trees are home to an array of nesting, breeding and migratory birds. When mangrove forests are cleared valuable habitat is lost, threatening the survival of myriad species.

But that's not the whole story. The forests are also a potential source of undiscovered biological materials that could benefit mankind, such as antibacterial compounds and pest-resistant genes, which are also lost when coastal areas are cleared.

References

- Estuary, science, britannica.com, Retrieved 14 June, 2019
- Chen, Si; Torres, Raymond (21 March 2012). "Effects of Geomorphology on the Distribution of Metal Abundance in Salt Marsh Sediment". Estuaries and Coasts. 35 (4): 1018–1027. Doi:10.1007/s12237-012-9494-y
- Conceptual-diagram-of-estuary-habitat-in-the-Fitzroy-region-major-control-variable, Len-mckenzie, profile: researchgate.net, Retrieved 15 July, 2019
- Te Ara – The Encyclopedia of New Zealand (2005–2010). Plants of the Estuary. Retrieved 15 March 2010, from http://www.teara.govt.nz/en/estuaries/3. ISSN 1559-2723
- Ocean, science: britannica.com, Retrieved 16 August, 2019
- Intertidal-zone, coastal-marine, ecosystems, our-environment, education: crd.bc.ca, Retrieved 17 January, 2019
- Corals-and-coral-reefs, corals-and-coral-reefs, ocean-life: ocean.si.edu, Retrieved 18 February, 2019
- Lagoon-geography, science: britannica.com, Retrieved 19 March, 2019
- Reasons-to-protect-mangrove-forests-for-the-future, agenda: weforum.org, Retrieved 20 April, 2019

Understanding Marine Life

3

- Marine Microorganisms
- Marine Flora
- Marine Fauna
- Phytoplankton
- Zooplankton
- Bacterioplankton
- Marine Vertebrates
- Marine Reptile
- Seabird
- Marine Mammal
- Marine Worm
- Marine Invertebrates

Marine life consists of various plants, animals and other organisms that live in seas or oceans. It includes different types of marine microorganisms, marine fauna, marine flora, zooplanktons, bacterioplanktons, marine invertebrates, etc. This chapter has been carefully written to provide an easy understanding of these types of marine life.

The phrase 'marine life' refers to organisms that live in salt water. These can include a diverse array of plants, animals and microbes (tiny organisms) such as bacteria and archaea.

Marine Life are Adapted to Life in Saltwater

From the perspective of a land animal like us, the ocean can be a harsh environment. However, marine life are adapted to live in the ocean. Characteristics that help marine life thrive in a saltwater environment include the ability to regulate their salt intake or deal with large quantities of salt water, adaptations to obtain oxygen (e.g., a fish's gills), being able to withstand high water pressures, living in a place where they can get enough light, or being able to adjust to a lack of light. Animals and plants that live on the edge of the ocean, such as tide pool animals and plants, also need to deal with extremes in water temperature, sunlight, wind and waves.

Types of Marine Life

There is a huge diversity in marine species. Marine life can range from tiny, single-celled organisms to gigantic blue whales, which are the largest creatures on Earth. Below is a list of the major phyla, or taxonomic groups, of marine life.

Major Marine Phyla

The classification of marine organisms is always in flux. As scientists discover new species, learn more about the genetic makeup of organisms, and study museum specimens, they debate how organisms should be grouped.

Marine Animal Phyla

Some of the most well-known marine phyla are listed below. You can find a more complete list here. The marine phyla listed below are drawn from the list on the World Register of Marine Species.

- Annelida: This phylum contains segmented worms. An example of a segmented marine worm is the Christmas tree worm.
- Arthropoda: Arthropods have a segmented body, jointed legs and a hard exoskeleton for protection. This group includes lobsters and crabs.
- Chordata: Humans are in this phylum, which also includes marine mammals (cetaceans, pinnipeds, sirenians, sea otters, polar bears), fish, tunicates, seabirds and reptiles.
- Cnidaria: This is a diverse phylum of animals, many of whom have stinging structures called nematocysts. Animals in this phylum include corals, jellyfish, sea anemones, sea pens and hydras.
- Ctenophora: These are jelly-like animals, such as comb jellies, but they don't have stinging cells.
- Echinodermata: This is one of my favorite phylums. It includes such beautiful animals as sea stars, brittle stars, basket stars, sand dollars and sea urchins.
- Mollusca: This phylum includes snails, sea slugs, octopuses, squids, and bivalves such as clams, mussels and oysters.
- Porifera: This phylum includes sponges, which are living animals. They can be very colorful and come in a diverse array of shapes and sizes.

Marine Plant Phyla

There are also several phyla of marine plants. These include the Chlorophyta, or green algae, and the Rhodophyta, or red algae.

Marine Microorganisms

Microorganisms make up about 70% of the marine biomass. A microorganism, or microbe, is a microscopic organism too small to be recognised with the naked eye. It can be single-celled or multicellular. Microorganisms are diverse and include all bacteria and archaea, most protozoa such as algae, fungi and certain microscopic animals such as rotifers. Many macroscopic animals and plants have microscopic juvenile stages. Some microbiologists also classify viruses (and viroids) as microorganisms, but others consider these as nonliving.

Microbial mats are the earliest form of life on Earth for which there is good fossil evidence. The image shows a cyanobacterial-algal mat.

Stromatolites are formed from microbial mats as microbes slowly move upwards to avoid being smothered by sediment.

Microorganisms are crucial to nutrient recycling in ecosystems as they act as decomposers. Some microorganisms are pathogenic, causing disease and even death in plants and animals. As inhabitants of the largest environment on Earth, microbial marine systems drive changes in every global system. Microbes are responsible for virtually all the photosynthesis that occurs in the ocean, as well as the cycling of carbon, nitrogen, phosphorus and other nutrients and trace elements.

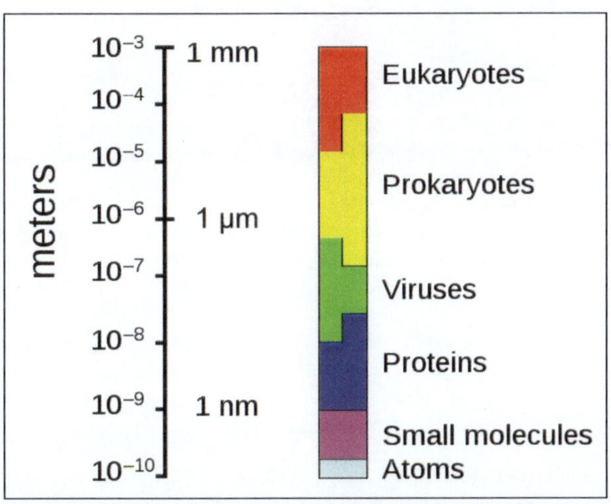

The range of sizes shown by prokaryotes (bacteria and archaea) and viruses relative to those of other organisms and biomolecules.

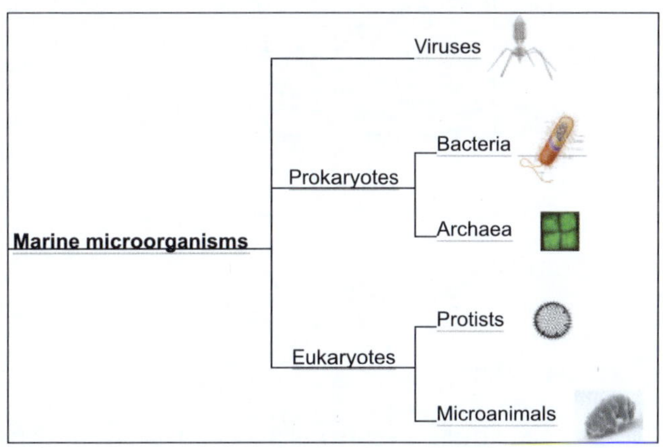

Microscopic life undersea is diverse and still poorly understood, such as for the role of viruses in marine ecosystems. Most marine viruses are bacteriophages, which are harmless to plants and animals, but are essential to the regulation of saltwater and freshwater ecosystems. They infect and destroy bacteria in aquatic microbial communities, and are the most important mechanism of recycling carbon in the marine environment. The organic molecules released from the dead bacterial cells stimulate fresh bacterial and algal growth. Viral activity may also contribute to the biological pump, the process whereby carbon is sequestered in the deep ocean.

A stream of airborne microorganisms circles the planet above weather systems but below commercial air lanes. Some peripatetic microorganisms are swept up from terrestrial dust storms, but most originate from marine microorganisms in sea spray. In 2018, scientists reported that hundreds of millions of viruses and tens of millions of bacteria are deposited daily on every square meter around the planet.

Microscopic organisms live throughout the biosphere. The mass of prokaryote microorganisms — which includes bacteria and archaea, but not the nucleated eukaryote microorganisms — may be as much as 0.8 trillion tons of carbon (of the total biosphere mass, estimated at between 1 and 4 trillion tons). Single-celled barophilic marine microbes have been found at a depth of 10,900 m (35,800 ft) in the Mariana Trench, the deepest spot in the Earth's oceans. Microorganisms live inside rocks 580 m (1,900 ft) below the sea floor under 2,590 m (8,500 ft) of ocean off the coast of the northwestern United States, as well as 2,400 m (7,900 ft; 1.5 mi) beneath the seabed off Japan. The greatest known temperature at which microbial life can exist is 122 °C (252 °F) (Methanopyrus kandleri). In 2014, scientists confirmed the existence of microorganisms living 800 m (2,600 ft) below the ice of Antarctica. According to one researcher, "You can find microbes everywhere — they're extremely adaptable to conditions, and survive wherever they are."

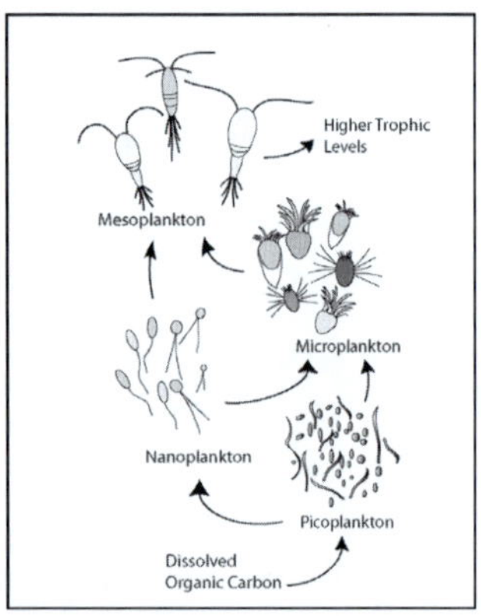

Marine microbial loop.

Sea spray containing marine microorganisms can be swept high into the atmosphere and may travel the globe before falling back to earth.

Marine Viruses

Viruses are small infectious agents that do not have their own metabolism and can replicate only inside the living cells of other organisms. Viruses can infect all types of life forms, from animals and plants to microorganisms, including bacteria and archaea. The linear size of the average virus is about one one-hundredth that of the average bacterium. Most viruses cannot be seen with an optical microscope so electron microscopes are used instead.

Viruses are found wherever there is life and have probably existed since living cells first evolved. The origin of viruses is unclear because they do not form fossils, so molecular techniques have been used to compare the DNA or RNA of viruses and are a useful means of investigating how they arose.

Viruses are now recognised as ancient and as having origins that pre-date the divergence of life into the three domains. But the origins of viruses in the evolutionary history of life are unclear: some may have evolved from plasmids—pieces of DNA that can move between cells—while others may have evolved from bacteria. In evolution, viruses are an important means of horizontal gene transfer, which increases genetic diversity.

Opinions differ on whether viruses are a form of life or organic structures that interact with living organisms. They are considered by some to be a life form, because they carry genetic material, reproduce by creating multiple copies of themselves through self-assembly, and evolve through natural selection. However they lack key characteristics such as a cellular structure generally considered necessary to count as life. Because they possess some but not all such qualities, viruses have been described as replicators and as "organisms at the edge of life".

Biological entities in the sea.Bacteriophages, often just called phages, are viruses that parasite bacteria and archaea. Marine phages parasite marine bacteria and archaea, such as cyanobacteria. They are a common and diverse group of viruses and are the most abundant biological entity in marine environments, because their hosts, bacteria, are typically the numerically dominant cellular life in the sea. Generally there are about 1 million to 10 million viruses in each mL of seawater, or about ten times more double-stranded DNA viruses than there are cellular organisms, although estimates of viral abundance in seawater can vary over a wide range. Tailed bacteriophages appear to dominate marine ecosystems in number and diversity of organisms. Bacteriophages belonging to the families Corticoviridae, Inoviridae and Microviridae are also known to infect diverse marine bacteria. Microorganisms make up about 70% of the marine biomass. It is estimated viruses kill 20% of this biomass each day and that there are 15 times as many viruses in the oceans as there are bacteria and archaea. Viruses are the main agents responsible for the rapid destruction of harmful algal blooms, which often kill other marine life. The number of viruses in the oceans decreases further offshore and deeper into the water, where there are fewer host organisms.

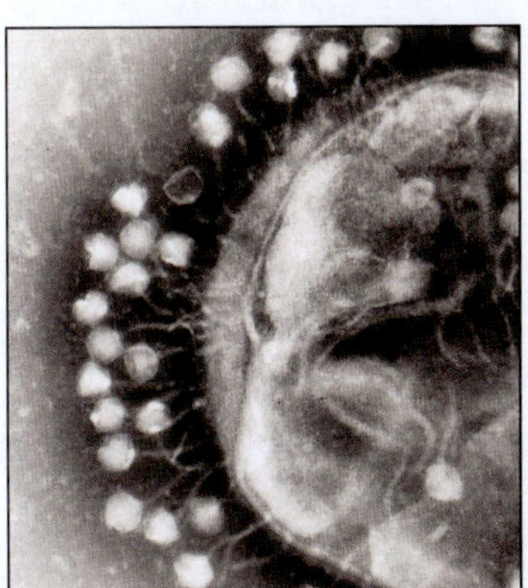

Multiple phages attached to a bacterial cell wall at 200,000x magnification.

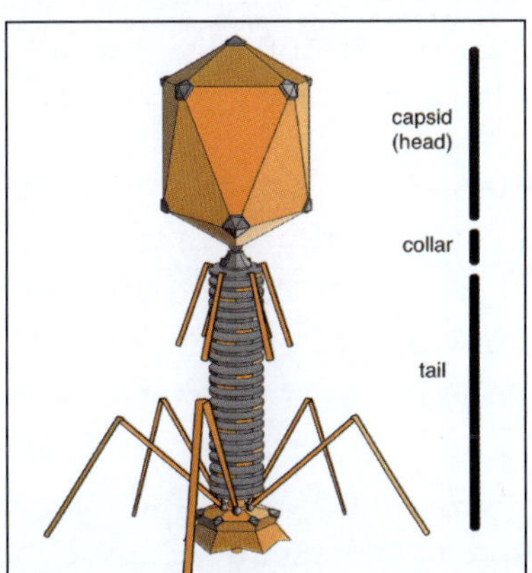

Diagram of a typical tailed phage.

There are also archaean viruses which replicate within archaea: these are double-stranded DNA viruses with unusual and sometimes unique shapes. These viruses have been studied in most detail in the thermophilic archaea, particularly the orders Sulfolobales and Thermoproteales.

Viruses are an important natural means of transferring genes between different species, which increases genetic diversity and drives evolution. It is thought that viruses played a central role in the early evolution, before the diversification of bacteria, archaea and eukaryotes, at the time of the last universal common ancestor of life on Earth. Viruses are still one of the largest reservoirs of unexplored genetic diversity on Earth.

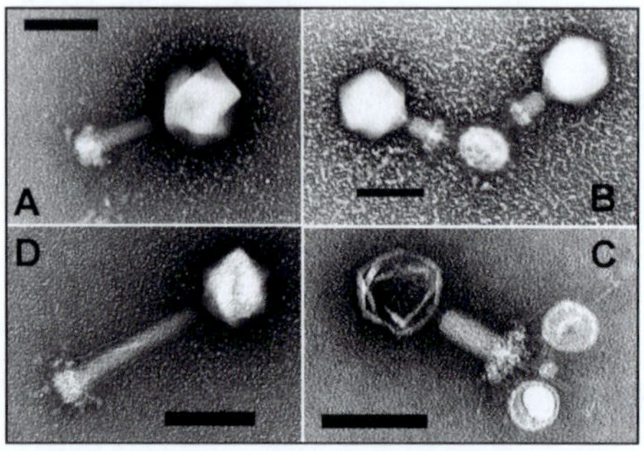

These are cyanophages, viruses that infect cyanobacteria (scale bars indicate 100 nm).

In terms of individual counts, tailed phage are the most abundant biological entities in the sea.

Marine Bacteria

Bacteria constitute a large domain of prokaryotic microorganisms. Typically a few micrometres in length, bacteria have a number of shapes, ranging from spheres to rods and spirals. Bacteria were among the first life forms to appear on Earth, and are present in most of its habitats. Bacteria inhabit soil, water, acidic hot springs, radioactive waste, and the deep portions of Earth's crust. Bacteria also live in symbiotic and parasitic relationships with plants and animals.

Once regarded as plants constituting the class Schizomycetes, bacteria are now classified as prokaryotes. Unlike cells of animals and other eukaryotes, bacterial cells do not contain a nucleus and rarely harbour membrane-bound organelles. Although the term bacteria traditionally included all prokaryotes, the scientific classification changed after the discovery in the 1990s that prokaryotes consist of two very different groups of organisms that evolved from an ancient common ancestor. These evolutionary domains are called Bacteria and Archaea.

The ancestors of modern bacteria were unicellular microorganisms that were the first forms of life to appear on Earth, about 4 billion years ago. For about 3 billion years, most organisms were microscopic, and bacteria and archaea were the dominant forms of life. Although bacterial fossils exist, such as stromatolites, their lack of distinctive morphology prevents them from being used to examine the history of bacterial evolution, or to date the time of origin of a particular bacterial species. However, gene sequences can be used to reconstruct the bacterial phylogeny, and these studies indicate that bacteria diverged first from the archaeal/eukaryotic lineage. Bacteria were also involved in the second great evolutionary divergence, that of the archaea and eukaryotes. Here, eukaryotes resulted from the entering of ancient bacteria into endosymbiotic associations with the ancestors of eukaryotic cells, which were themselves possibly related to the Archaea. This involved the engulfment by proto-eukaryotic cells of alphaproteobacterial symbionts to form either mitochondria or hydrogenosomes, which are still found in all known Eukarya. Later on, some eukaryotes that already contained mitochondria also engulfed cyanobacterial-like organisms. This led to the formation of chloroplasts in algae and plants. There are also some algae that originated from even later endosymbiotic events. Here, eukaryotes engulfed a eukaryotic algae that developed into a "second-generation" plastid. This is known as secondary endosymbiosis.

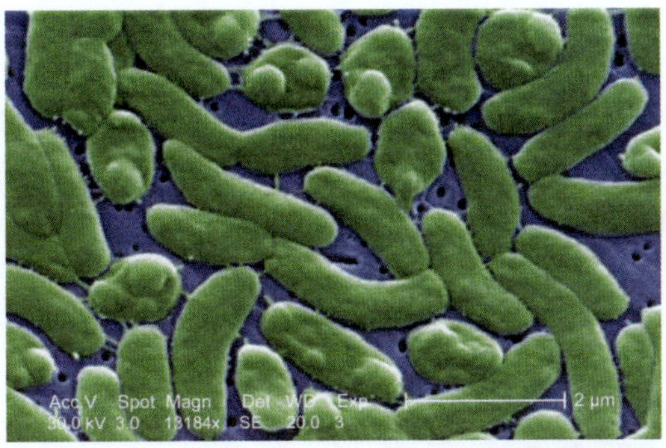

Vibrio vulnificus, a virulent bacterium found in estuaries and along coastal areas.

Cyanobacteria blooms can contain lethal cyanotoxins.

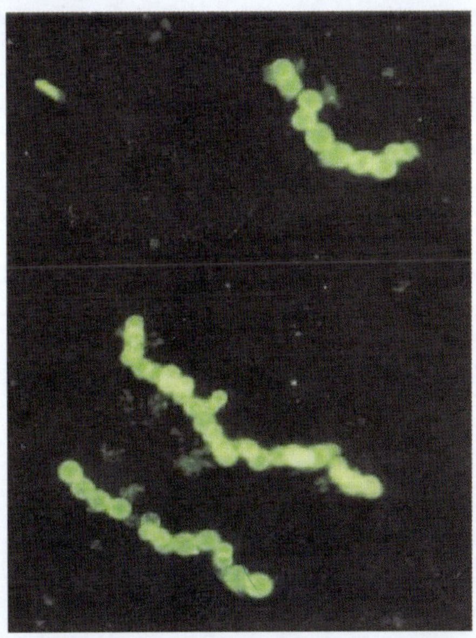
The marine Thiomargarita namibiensis, largest known bacterium.

The largest known bacterium, the marine Thiomargarita namibiensis, can be visible to the naked eye and sometimes attains 0.75 mm (750 µm).

Marine Archaea

The archaea constitute a domain and kingdom of single-celled microorganisms. These microbes are prokaryotes, meaning they have no cell nucleus or any other membrane-bound organelles in their cells.

Archaea were initially classified as bacteria, but this classification is outdated. Archaeal cells have unique properties separating them from the other two domains of life, Bacteria and Eukaryota.

The Archaea are further divided into multiple recognized phyla. Classification is difficult because the majority have not been isolated in the laboratory and have only been detected by analysis of their nucleic acids in samples from their environment.

Archaea and bacteria are generally similar in size and shape, although a few archaea have very strange shapes, such as the flat and square-shaped cells of Haloquadratum walsbyi. Despite this morphological similarity to bacteria, archaea possess genes and several metabolic pathways that are more closely related to those of eukaryotes, notably the enzymes involved in transcription and translation. Other aspects of archaeal biochemistry are unique, such as their reliance on ether lipids in their cell membranes, such as archaeols. Archaea use more energy sources than eukaryotes: these range from organic compounds, such as sugars, to ammonia, metal ions or even hydrogen gas. Salt-tolerant archaea (the Haloarchaea) use sunlight as an energy source, and other species of archaea fix carbon; however, unlike plants and cyanobacteria, no known species of archaea does both. Archaea reproduce asexually by binary fission, fragmentation, or budding; unlike bacteria and eukaryotes, no known species forms spores.

Archaea were initially viewed as extremophiles living in harsh environments, such as the yellow archaea pictured here in a hot spring, but they have since been found in a much broader range of habitats.

Archaea are particularly numerous in the oceans, and the archaea in plankton may be one of the most abundant groups of organisms on the planet. Archaea are a major part of Earth's life and may play roles in both the carbon cycle and the nitrogen cycle.

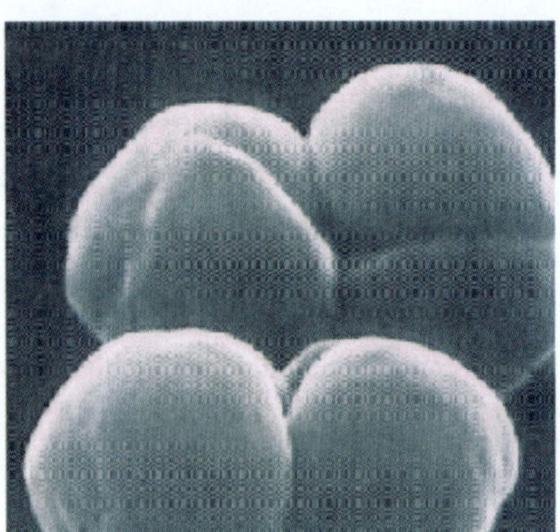

Methanosarcina barkeri, a marine archaea that produces methane.

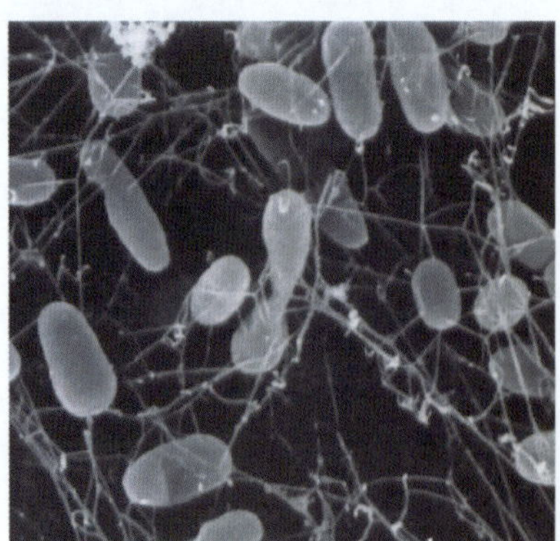

Thermophiles, such as Pyrolobus fumarii, survive well over 100 °C.

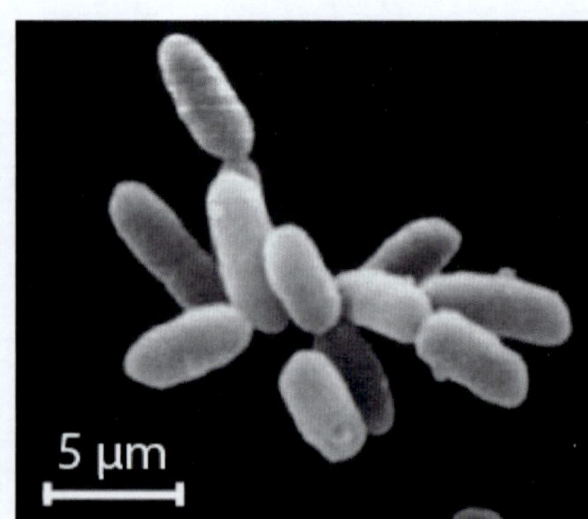

Halobacteria, found in water near saturated with salt, are now recognised as archaea.

Marine Protists

All living organisms can be grouped as either prokaryotes or eukaryotes. Life originated as single-celled prokaryotes and later evolved into the more complex eukaryotes. In contrast to prokaryotic cells, eukaryotic cells are highly organised. Prokaryotes are the bacteria and archaea, while eukaryotes are the other life forms — plants, animals, fungi and protists. Plants, animals or fungi are usually multi-celled. Protists are eukaryotes that cannot be classified as plants, animals or fungi.

Protists fall into three groups, according to whether their nutrition is animal-like, plant-like, or fungus-like:

- Algae are the plant-like protists that are autotrophs that make their own food using photosynthesis.
- Protozoa are the animal-like protists that are heterotrophs that ingest or absorb their food.
- Slime moulds and water moulds are the fungus-like protists that are also heterotrophs like protozoa.

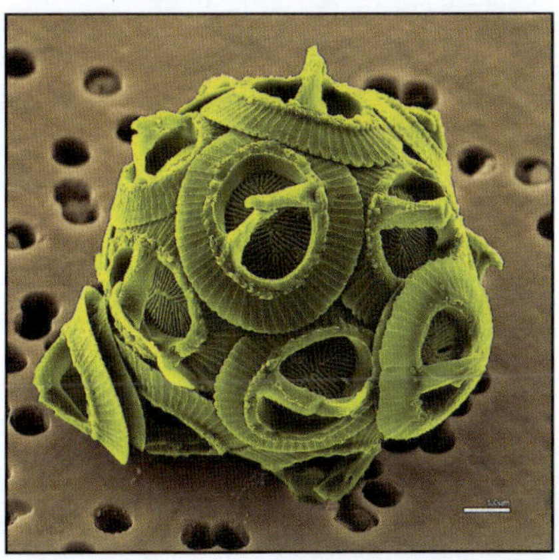

Single-celled alga, Gephyrocapsa oceanica.

Red algae, Cyanidium sp.

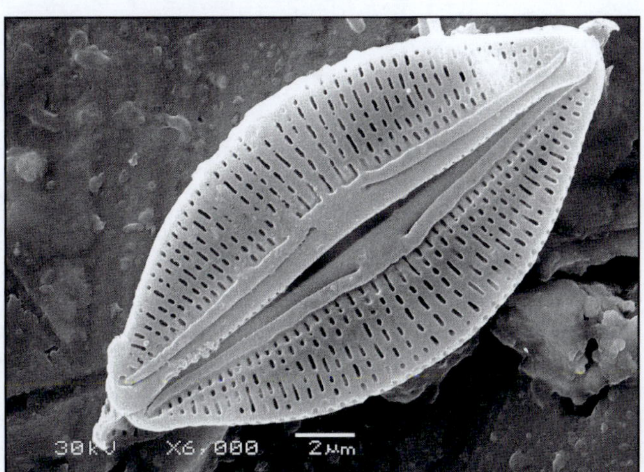

Silicified cell wall of a diatom Amphora sp.

Protists include amoebae, ciliates, red algae, euglena, slime molds, and phytoplankton such as diatoms and dinoflagellates. Several algae species are multicellular protists, and some marine slime molds have unique life cycles that involve switching between unicellular, colonial, and multicellular forms. Protists are a highly diverse group of organisms currently organised into 18 phyla, but they are not easy to classify. Studies have shown a high protist diversity exists in oceans, deep sea-vents and river sediments, which suggests a large number of eukaryotic microbial communities have yet to be discovered.

Most protists are single-celled and microscopic. But some single-celled marine protists are macroscopic. Other marine protist are neither single-celled nor microscopic, such as seaweed.

The unicellular bubble algae lives in tidal zones. It can have a 4 cm diameter.

The single-cell xenophyophore lives in abyssal zones. It has a giant shell up to 20 cm across.

Marine Microanimals

As juveniles, animals develop from microscopic stages, which can include spores, eggs and larvae. At least one microscopic animal group, the parasitic cnidarian Myxozoa, is unicellular in its adult form, and includes marine species. Other adult marine microanimals are multicellular. Microscopic adult arthropods are more commonly found inland in freshwater, but there are marine species as well. Microscopic adult marine crustaceans include some copepods, cladocera and water bears. Some marine nematodes and rotifers are also too small to be seen with the naked eye, as are many loricifera, including the recently discovered anaerobic species that spend their lives in an anoxic environment. Copepods contribute more to the secondary productivity and carbon sink of the world oceans than any other group of organisms.

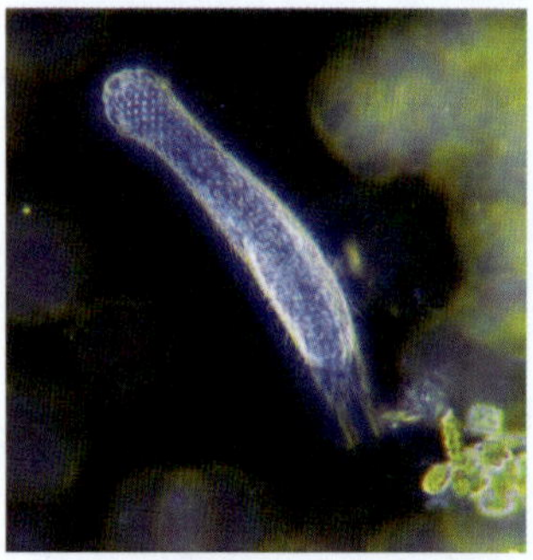

Darkfield photo of a gastrotrich, which form a phylum of worm-like animals living mostly between particles of sediment.

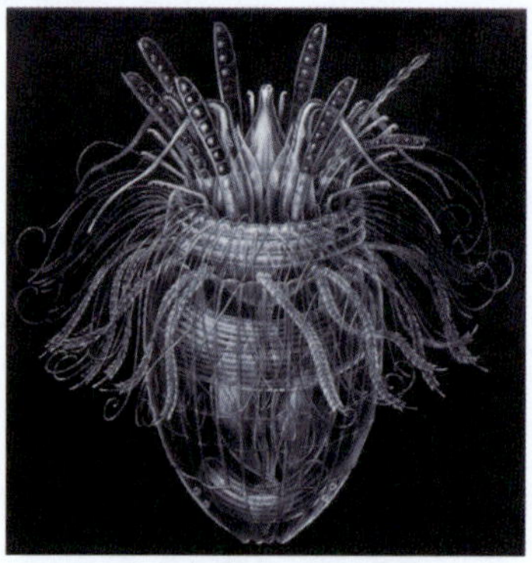

Armoured Pliciloricus enigmaticus of the phylum Loricifera live in the spaces between marine gravel.

Marine Flora

There are thousands of different types of plants found in the ocean, and all of them are affected by the amount of available sunlight, salinity levels, and the temperature of the water. Ocean plants are rarely talked about, but with 70 percent of the world covered in water, they play a vital role in oxygenating oceans around the world, protecting certain species of aquatic animals from predators and outside elements (including human activities), and helping animals thrive by being their primary food source.

Common Types of Ocean Plants

There are basically two categories of plants living in our ocean's waters: those that float freely through the water, and those that are directly rooted into the ocean floor. Floating plants can be spotted near the surface of the water. Rooted plants, on the other hand, must acquire sunlight to survive, so they are often found in shallow waters near the shore.

The four common types of floating and rooted plants that live in the ocean:

Kelp

Kelp beds are commonly found throughout colder ocean waters. As a form of algae, kelp grows along rocky coastlines at depths of 20 to 90 feet, clinging to rocks with their root-like structure or staying afloat through the use of tiny flotation bulbs. This oceanic plant thrives on sunlight and it can often grow up to 250 feet in length, making it the largest marine plant in the world.

Red Algae

Red algae have populated the warm, tropical waters of the world for more than 500 million years. It gets its unique coloration from phycoerythrin, a pigment that absorbs blue light and reflects red light. Similar to kelp, red algae is rich in proteins and vitamins, making it a popular food source for countries throughout Asia.

You'll also find a unique type of red algae, the coralline algae, which are made of very fine filaments and are hard as the rocks that they grow on to support the structure of coral reefs.

Seagrass

Seagrass is a rooted plant that requires a lot of light to thrive along the shallow waters of our world's coastlines. There are many common types of sea grass, including manatee grass, turtle grass, and shoal grass—all of which are varieties of a land plant that has evolved over the years in order to survive in salt water.

It is also a popular food source for crabs and lobsters throughout the temperate marsh waters where it commonly grows. Not only does seagrass greatly contribute to the task of oxygenating the oceans, but it also releases pollen from its flowers to pollinate other ocean plants, provides a protective habitat for small marine life, and serves as food for some species of reef fish.

This makes seagrass one of the most important plants found in the ocean.

Sargassum

Sargassum, also known as gulfweed or sea holly, is a floating ocean plant that plays an integral role in the early lives of many marine species. It is characterized by its brownish-green coloration, dense, bushy fronds, and berry-like gas-filled bladders that keep it afloat.

The bladders' floating nature keeps the plant up at the surface of the water, better enabling photosynthesis where it can be closer to the sun and allowing juvenile sea turtles to turn to it for food and shelter as they mature and eventually make the long trip back home.

Sargassum is so abundant in a gyre of the North Atlantic that it has been dubbed the Sargasso Sea. Some marine species found within the Sargasso Sea are found nowhere else in the world!

Ocean plants are part of some of the most important ecosystems on the planet, so it is imperative that we all do our best to protect and take care of our marine gardens.

Seagrass

Seagrasses are found in shallow salty and brackish waters in many parts of the world, from the tropics to the Arctic Circle. Seagrasses are so-named because most species have long green, grass-like leaves. They are often confused with seaweeds, but are actually more closely related to the flowering plants that you see on land. Seagrasses have roots, stems and leaves, and produce flowers and seeds. They evolved around 100 million years ago, and today there are approximately 72 different seagrass species that belong to four major groups. Seagrasses can form dense underwater meadows, some of which are large enough to be seen from space. Although they often receive little attention, they are one of the most productive ecosystems in the world. Seagrasses provide shelter and food to an incredibly diverse community of animals, from tiny invertebrates to large fish, crabs, turtles, marine mammals and birds. Seagrasses provide many important services to people as well, but many seagrasses meadows have been lost because of human activities. Work is ongoing around the world to restore these important ecosystems.

A Plant, not a Seaweed

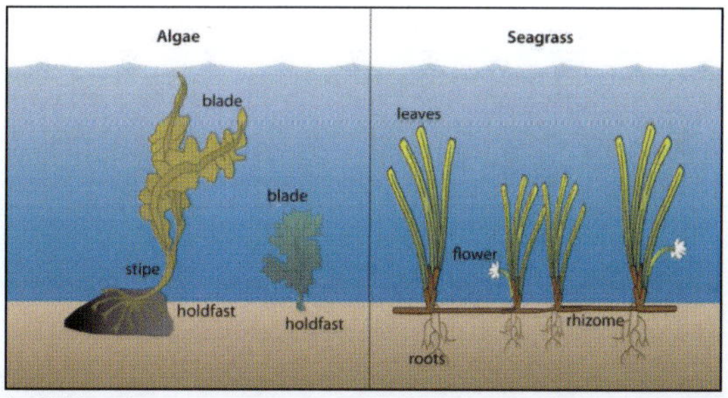

Algae or "seaweeds" (left) differ from seagrasses (right) in several ways. Algae on the seafloor have a holdfast and transport nutrients through the body by diffusion, while seagrasses are flowering vascular plants with roots and an internal transport system.

Even though seagrasses and seaweeds look superficially similar, they are very different organisms. Seagrasses belong to a group of plants called monocotyledons that include grasses, lilies and palms. Like their relatives, seagrasses have leaves, roots and veins, and produce flowers and seeds. Chloroplasts in their tissues use the sun's energy to convert carbon dioxide and water into sugar and oxygen for growth through the process of photosynthesis. Veins transport nutrients and water throughout the plant, and have little air pockets called lacunae that help keep the leaves buoyant and exchange oxygen and carbon dioxide throughout the plant. Like other flowering plants, their roots can absorb nutrients. Unlike flowering plants on land, however, they lack stomata—the tiny pores on leaves that open and close to control water and gas exchange. Instead, they have a thin cuticle layer, which allows gasses and nutrients to diffuse directly into and out of the leaves from the water. The roots and rhizomes (thicker horizontal stems) of seagrasses extend into the sediment of the seafloor and are used to store and absorb nutrients, as well as anchor the plants. In contrast, seaweeds (algae) are much simpler organisms. They have no flowers or veins, and their holdfasts simply attach to the bottom and are generally not specialized to take in nutrients. Scientists are studying what genes were lost and which were regained as seagrasses evolved from algae in the sea to plants on land, and then transitioned back to the sea. The entire genome of one seagrass, the eelgrass Zostera marina, was sequenced in 2016, helping us understand how these plants adapted to life in the sea, how they may respond to climate warming, and the evolution of salt tolerance in crop plants.

Where are Seagrasses Found?

Seagrasses grow in salty and brackish (semi-salty) waters around the world, typically along gently sloping, protected coastlines. Because they depend on light for photosynthesis, they are most commonly found in shallow depths where light levels are high. Many seagrass species live in depths of 3 to 9 feet (1 to 3 meters), but the deepest growing seagrass (Halophila decipiens) has been found at depths of 190 feet (58 meters). While most coastal regions are dominated by one or a few seagrass species, regions in the tropical waters of the Indian and western Pacific oceans have the highest seagrass diversity with as many as 14 species growing together. Antarctica is the only continent without seagrasses.

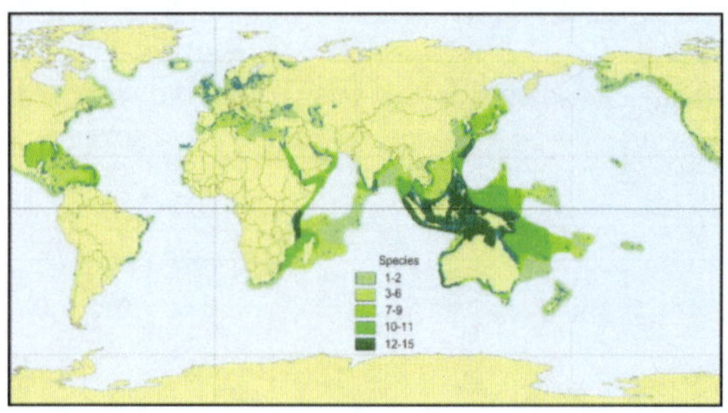

Seagrasses are found across the world, from the tropics to the Arctic. Shades of green indicate the number of species reported for a given area. The darker shades of green indicate more species are present.

Growth and Reproduction

Seagrasses grow both vertically and horizontally—their blades reach upwards and their roots down and sideways—to capture sunlight and nutrients from the water and sediment. They spread by two methods: asexual clonal growth and sexual reproduction.

- Asexual Clonal Growth: Similar to grasses on land, seagrass shoots are connected underground by a network of large root-like structures called rhizomes. The rhizomes can spread under the sediment and send up new shoots. When this happens, many stems within the same meadow can actually be part of the same plant and will have the same genetic code—which is why it is called clonal growth. In fact, the oldest known plant is a clone of the Mediterranean seagrass Posidonia oceanica, which may be up to 200,000 years old, dating back to the ice ages of the late Pleistocene. In some seagrass species, a meadow can develop from a single plant in less than a year, while in slow-growing species like Posidonia it can take hundreds of years.

- Sexual Reproduction: Seagrasses reproduce sexually like terrestrial grasses, but pollination for seagrasses is completed with the help of water. Male seagrass flowers release pollen from structures called stamens into the water. Seagrasses produce the longest pollen grains on the planet (up to 5mm long compared to under 0.1mm for land plants typically), and this pollen often collects into stringy clumps. The clumps are moved by currents until they land on the pistil of a female flower and fertilization takes place. There is also evidence that small invertebrates, such as amphipods (tiny shrimp-like crustaceans) and polychaetes (marine worms), feed on the pollen of one seagrass (Thalassia testudinum), which could help to fertilize the flowers in a way similar to how insects pollinate flowers on land.

Self-pollination happens in some grass species, which can reduce genetic variation. Individual seagrass plants avoid this by producing only male or female flowers, or by producing the male and female flowers at different times. Just like land grasses, fertilized seagrass flowers develop seeds. Seagrass seeds are neutrally buoyant and can float many miles before they settle onto the soft seafloor and germinate to form a new plant. A few seagrass species such as the surfgrass Phylospadix can settle and live on rocky shores. Animals that eat seagrass seeds—including fish and turtles—may incidentally aid with their dispersal and germination if the seeds pass through their digestive tracks and remain viable.

Biodiversity

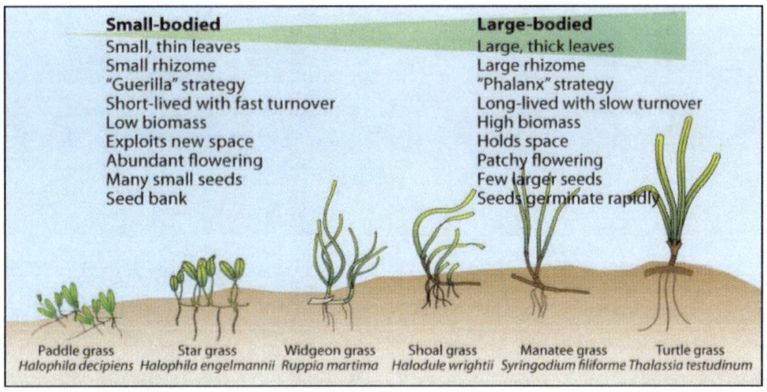

The 72 species of seagrasses are commonly divided into four main groups: Zosteraceae, Hydrocharitaceae, Posidoniaceae and Cymodoceaceae. Their common names, like eelgrass, turtle grass, tape grass, shoal grass, and spoon grass, reflect their many shapes and sizes and roles in marine ecosystems. Seagrasses range from species with long flat blades that look like ribbons to fern or paddle-shaped leaves, cylindrical or spaghetti blades, or branching shoots. The tallest seagrass species—Zostera caulescens—was found growing to 35 feet (7 meters) in Japan. Some seagrass species are quick growing while others grow much more slowly. These distinct structures and growth forms affect how seagrasses influence their environment and what species live in the habitats they create.

Ecosystem Benefits

Seagrasses are often called foundation plant species or ecosystem engineers because they modify their environments to create unique habitats. These modifications not only make coastal habitats more suitable for the seagrasses themselves, but also have important effects on other animals and provide ecological functions and a variety of services for humans.

Seagrasses have been used by humans for over 10,000 years. They've been used to fertilize fields, insulate houses, weave furniture, thatch roofs, make bandages, and fill mattresses and even car seats. But it's what they do in their native habitat that has the biggest benefits for humans and the ocean. Seagrasses support commercial fisheries and biodiversity, clean the surrounding water and help take carbon dioxide out of the atmosphere. Because of these benefits, seagrasses are believed to be the third most valuable ecosystem in the world (only preceded by estuaries and wetlands). One hectare of seagrass (about two football fields) is estimated to be worth over $19,000 per year, making them one of the most valuable ecosystems on the planet.

Key Services

Modification of the Physical Environment

An Australian Giant Cuttlefish (Sepia apama) crosses a seagrass bed.

Seagrasses are known as the "lungs of the sea" because one square meter of seagrass can generate 10 liters of oxygen every day through photosynthesis. Seagrass leaves also absorb nutrients and slow the flow of water, capturing sand, dirt and silt particles. Their roots trap and stabilize the sediment, which not only helps improve water clarity and quality, but also reduces erosion and buffers coastlines against storms. Seagrasses can further improve water quality by absorbing nutrients in runoff from the land. In nutrient poor regions, the seagrass plants themselves help nutrient cycling by taking up nutrients from the soil and releasing them into the water through their leaves, acting as a nutrient pump.

Creation of Living Habitat

Seagrasses don't just provide shelter for free-swimming animals, but also are a habitat for non-moving organisms, such as these sea anemones.

Seagrasses are often called nursery habitats because the leafy underwater canopy they create provides shelter for small invertebrates (like crabs and shrimp and other types of crustaceans), small fish and juveniles of larger fish species. Many species of algae and microalgae (such as diatoms), bacteria and invertebrates grow as "epiphytes" directly on living seagrass leaves, much like lichens and Spanish moss grow on trees. Other invertebrates grow nestled between the blades or in the sediments—such as sponges, clams, polychaete worms and sea anemones. The accumulation of smaller organisms amongst and on the seagrass blades, as well as the seagrass itself, attracts bigger animals. As a result, seagrasses can be home to many types of fish, sharks, turtles, marine mammals (dugongs and manatees), mollusks (octopus, squid, cuttlefish, snails, bivalves), sponges, crustaceans (shrimp, crabs, copepods, isopods and amphipods) polychaete worms, sea urchins and sea anemones—and the list goes on.

Some of these organisms are permanent residents in seagrass meadows, while others are temporary visitors. A single acre of seagrass can support upwards of 40,000 fish and 50 million small invertebrates, and there are often tens to hundreds more animals in a seagrass bed compared to adjacent bare sandy areas. Hundreds of species live in the seagrass near the Smithsonian Marine Station at Fort Pierce in Florida. A number of the species that depend on seagrasses are important for commercial and recreational fisheries. In fact, in all regions of the world fishermen will specifically seek out seagrass beds for their abundance of fish. It is because of the wide variety of different species that live amongst the grasses that seagrass beds often form important "biodiversity hotspots." Not only do seagrasses support a diversity of marine life, but populations of a given seagrass species can themselves be very genetically diverse and this diversity itself is linked to higher animal abundances. Understanding how seagrass genotypic diversity does this is an active area of research.

Foundation of Coastal Food Webs

Adult green sea turtles spend most of their time grazing in seagrass meadows.

Seagrass beds are important feeding grounds for thousands of species around the world, and they support this diverse food web in three different ways. Some organisms—primarily large grazers like manatees, dugongs, green sea turtles and geese—eat the living leaves directly, and seagrass forms a major component of their diets. For example, an adult dugong eats about 64 to 88 pounds (28 to 40 kg) of seagrass a day, while an adult green sea turtle can eat about 4.5 pounds (2 kg) per day. Many of these large grazers are endangered, in large part because of habitat destruction and hunting, but once they were very common. It's estimated that before Europeans settled the Americas in the 1400's, the number of green sea turtles supported by seagrass meadows was 15 to 20 times the number and biomass of large hooved animals in the Serengeti Desert alive today. These abundant large grazers probably kept seagrass meadows cropped short like a putting green.

The epiphytic organisms growing on the surface of the seagrass blades provide other sources of food. Some epiphytic bacteria can extract nitrogen from the environment and make it available to larger animals. Small invertebrate mesograzers, such as crustaceans and snails, feed on epiphytes, and in doing so can help keep the seagrass clean, acting as mutualistic partners (or housekeepers) that promote seagrass growth. They are in turn consumed by larger crustaceans, fish and birds and are important links in the coastal food web. But, this partnership isn't always positive.

Occasionally when some mesograzer species are at very high densities they can create thick masses of mucus and sediment tubes that block light to the seagrass leaves, and they can even eat the seagrass directly.

Dead seagrass leaves also play an important role in coastal ecosystems. When the leaves die, they decay on the sediment or are washed onto the beach, supporting a diverse community of decomposers that thrive on rotting material. Some of these living and dead seagrass blades are also washed to other areas of the ocean, feeding organisms in ecosystems as far as the deep sea.

Blue Carbon

Seagrasses are capable of capturing and storing a large amount of carbon from the atmosphere. Similar to how trees take carbon from the air to build their trunks, seagrasses take carbon from the water to build their leaves and roots. As parts of the seagrass plants and associated organisms die and decay, they can collect on the seafloor and become buried, trapped in the sediment. It has been estimated that in this way the world's seagrass meadows can capture up to 83 million metric tons of carbon each year. The carbon stored in sediments from coastal ecosystems including seagrass meadows, mangrove forests and salt marshes is known as "blue carbon" because it is stored in the sea. While seagrasses occupy only 0.1 percent of the total ocean floor, they are estimated to be responsible for up to 11 percent of the organic carbon buried in the ocean. One acre of seagrass can sequester 740 pounds of carbon per year (83 g carbon per square meter per year), the same amount emitted by a car traveling around 3,860 miles (6,212 km).

Atmospheric carbon is captured by coastal mangroves, seagrasses and salt marshes at a rate five times faster than tropical forests.

Threats and Conservation

Unfortunately, seagrasses are in trouble. Seagrass coverage is being lost globally at a rate of 1.5 percent per year. That amounts to about 2 football fields of seagrass lost each hour. It's estimated that 29 percent of seagrass meadows have died off in the past century. In a 2011 assessment, nearly one quarter of all seagrass species for which information was adequate to judge were threatened (endangered or vulnerable) or near threatened using the International Union for the Conservation of Nature (IUCN) Red List criteria. This is especially worrying because seagrass losses are projected to have severe impacts on marine biodiversity, the health of other marine ecosystems, and on human livelihoods. Additionally, some threatened marine species such as sea turtles and marine mammals live in seagrass habitats and rely on them for food. For every seagrass species there is on average more than one associated threatened marine species. In fact, the only marine plant listed as endangered in the United States is a seagrass (Halophila johnsonii) found in Florida.

Threats to Seagrasses

A strain of Caulerpa taxifolia escaped aquariums and has spread widely in the Mediterranean, replacing native plants and depriving marine life of food and habitat.

Seagrasses are vulnerable to physical disturbances, such as wind-driven waves and storms. Some animals, such as skates and rays, disturb the rhizomes and roots of seagrasses, ripping up the seagrass as they forage for buried clams and other invertebrates. However, the direct and indirect effects of human activities account for most losses of seagrass beds in recent decades. Some fast growing seagrass meadows are able to rebound from disturbances, but many grow slowly over the course of centuries and are likely to be slow to recover and are thus most vulnerable.

Nutrients, such as those from fertilizers and pollution, wash off the land and into the water, causing algal blooms that block sunlight necessary for seagrass growth. Sediment washing into the water from agriculture and land development can also damage seagrass beds by both smothering the seagrass and blocking sunlight. Similarly, dredging can both directly remove seagrass plants and cause lower light levels because of increased amounts of sediments in the water. Boat anchors and propellers can leave "scars" in a seagrass bed–killing sections of the seagrass and fragmenting the habitat. This fragmentation of seagrass beds can increase erosion around the edges, as well as influence animal use and movement within the seagrass bed.

Disease has also devastated seagrasses. In the early 1930s, a large die-off of up to 90 percent of all eelgrass (Zostera marina) growing in temperate North America was attributed to a "wasting disease". This die-off was so severe that a small snail specialized to live on eelgrass went extinct as a result. The disease was caused by the slime mold-like protist, Labyrinthula zosterae, which also ravaged eelgrass populations in Europe. This disease still affects eelgrass populations in the Atlantic and has contributed to some recent losses, though none as catastrophic as in the 1930s. Eelgrass leaves that are weak or stressed are more susceptible to the disease, developing brown spots and lesions that reduce the plant's ability to photosynthesize, eventually killing the plant. Healthy plants are thought to be resistant to the disease, indicating importance of reducing other stressors like pollution. Lower seawater salinity may also increase susceptibility to the Labyrinthula pathogen.

Episodes of warm seawater temperatures can also damage seagrasses. Temperature affects how enzymes and metabolism work, influencing how organisms grow. Rising water temperatures tend to increase rates of seagrass respiration (using up oxygen) faster than rates of photosynthesis (producing oxygen), which makes them more susceptible to grazing by herbivores. Increased temperature also increases seagrass light requirements, influences how quickly seagrasses can take up nutrients in their environment, and can make seagrasses more susceptible to disease. Large eelgrass declines have been observed in the Chesapeake Bay in years in which water temperatures have persisted for several days above 30 °C (86 °F), the thermal limit for this species.

Removal of fish can also lead to seagrass death by disrupting important components of the food web. When large predators are removed, intermediate predators can become more abundant, and they in turn cause the decline of the smaller organisms that keep the blades of the seagrasses clean. This has been observed most strikingly in the Baltic sea with the disappearance of cod due to overfishing and corresponding increases in smaller fishes and crustaceans which limited epiphyte-grazing invertebrates, resulting in seagrass decline.

In addition to the small epiphytic algae, larger algae also compete with seagrasses, and introduced invasive seaweed species can displace native seagrass species. One important example is the invasion of Caulerpa taxifolia, a seaweed nicknamed "the killer algae." Released into the Mediterranean in the 1980s from aquaria, by 2000 it covered more than 131 square kilometers (50 square miles) of the Mediterranean coastline, overgrowing and replacing the native Neptune seagrass (Posidonia oceanica) and reducing the ecosystem's biodiversity. Since then, invasive Caulerpa has been found in California and southwestern Australia where eradication programs are in place to prevent its spread.

Protecting and Restoring Seagrass Beds

Neptune grass (Posidonia oceanica) is a slow-growing and long-lived seagrass native to the Mediterranean.

Most management that protects seagrasses focuses on maintaining their biodiversity and the services these habitats provide for humans and ecosystems. There is no international legislation for seagrasses, and so protection typically occurs by local and regional agencies. Actions taken to help seagrasses include limiting damaging practices such as excessive trawling and dredging, runoff pollution and harmful fishing practices (such as dynamite or cyanide fishing).

There are also attempts to rebuild and restore seagrass beds, often by planting seeds or seedlings grown in aquaria, or transplanting adult seagrasses from other healthy meadows. Some of the most successful restoration stories come from the Chesapeake Bay and coastal Virginia in the Eastern United States where, through 2014, the Virginia Institute of Marine Science has seeded 456 acres with 7.65 million seagrass seeds. As of 2015, the seagrass Zostera marina has increased from these seeded plots to cover 6,195 acres. Seagrass restoration in Tampa Bay, Florida, has also experienced important success including improvements in water quality and the associated fish community. For restoration to work, it is critical that the causes of the original decline in seagrasses have been eliminated.

However, seagrass populations globally are still in trouble. Some simple steps everyone can take to help seagrasses and other marine habitats include: don't litter, limit the amount of fertilizer and pesticides you use, don't dump anything hazardous down the drain, be careful when boating by going slow and avoiding shallow areas, and support local conservation efforts.

Marine Fauna

Marine fauna are highly diverse and range in size from microscopic zooplankton to the blue whale. Marine fauna that live their entire life in the ocean include sharks, whales, dolphins, dugongs, sea snakes, most fish and crustaceans. Marine fauna are also animals that either leave or enter the ocean for breeding or resting purposes, such as turtles, seals and sea lions, penguins, and crabs. Other animals such as seabirds can also be considered as marine fauna as they rely on fish and other marine life for food. While animals such as sponges and corals that are attached to the seabed are also marine fauna, they are typically considered under the environmental factor of Benthic Communities and Habitats.

Environmental Values of Marine Fauna and their Significance

All species of marine fauna are important and are part of the broader marine and coastal ecosystem. However some of these species, or even groups of species, have critical roles to play in maintaining key ecological functions and processes within the ecosystem. For some species, the number of individuals have declined to levels where they are now considered threatened under State legislation. Other species are of social, cultural and/or economic significance to Western Australians.

These include commercially and recreationally important fish as well as iconic species important to local or regional communities such as turtles and dugongs.

Issues Related to Marine Fauna

The following issues are matters that are commonly encountered by the EPA due to the nature of proposals that are referred to it. Background on these issues is provided here to help proponents and the community engage with EIA. This issues section will be updated from time to time to reflect new issues as they arise in referrals and EIA.

Direct Impacts on Marine Fauna

Direct impacts on marine fauna include vessel strike or entanglement in equipment such as dredges and aquaculture cages. These impacts can be severe, but are localised and relatively easy to monitor and measure. As such they are amenable to adaptive management actions to avoid and minimise those impacts.

Indirect Impacts on Marine Fauna

Most indirect effects on marine fauna result from changing the characteristics of the marine and coastal environment. These can result from changes to chemical and physical properties such as temperature, pH, hydrodynamic processes, and water clarity. Other effects, such as those from underwater noise, can result in behavioral impacts that are more subtle and occur far from the activity in question. While some of these impacts will also be considered under other environmental factors, the consequent impacts on marine fauna can be much more difficult to quantify and manage. They may result in temporary changes during the construction phase of a proposal, or in long-term or permanent changes during the operational phases. Examples of these types of proposals include seismic surveys, pile driving, waste from discharges, installation of coastal and marine infrastructure, enclosure of marine and coastal waters in marinas and ports, and increased vessel movements. The focus of the EPA in assessing proposals is on both minimising emissions and maintaining the characteristics of the marine and coastal environment that support marine fauna. Consideration and assessment of indirect impacts, particularly with regard to the potential to changes to key environmental parameters, should be given appropriate focus during EIA for marine fauna.

Impacts to Critical Habitats and during Key Ecological Windows

Marine fauna rely on a range of ecological conditions, physical and chemical properties of the marine and coastal environment and specific habitats during key stages of their lifecycle. Turtles, for example, nest on beaches that meet specific physical characteristics regarding beach profile, and come ashore during specific times of the year when wave energy is low. Many fish and invertebrate species rely on specific habitats such as seagrass meadows to provide protection and food resources for juveniles. Other species migrate to specific areas to spawn, relying on specific hydrodynamic conditions and physical properties such as water temperature as signals to prompt aggregation and breeding. Undertaking development activities within these key areas and/or during these key times has the potential to have a significant impact on these species. As a result, avoiding these locations and times is important. The focus of the EPA is ensuring that these critical habitats and key ecological windows have been appropriately identified and impacts minimised through avoidance where possible and as much as practicable.

Changing Climate

Biological diversity is vulnerable to a changing climate. The scale, rate and nature of the projected change, and the interactions between climate change and other threatening processes, have the potential to overwhelm the capacity of current ecosystems to adapt. From a marine perspective, the consequences of climate change such as increasing storms and wave energy, marine heatwaves, and ocean acidification are likely to significantly affect marine fauna. While some species may be able to adapt and others may be able to shift their ranges, this may not be the case for all species. For the purposes of EIA, the EPA is concerned with proposal specific impacts that, when considered in combination with climate change, are likely to exacerbate changes to biological diversity and ecological integrity.

State of the Science and Knowledge

The EPA recognises that the current state of scientific knowledge for marine fauna is highly variable. The size and remote nature of much of Western Australia, combined with a high level of biological diversity and a lack of full knowledge, is a key challenge to impact assessment for marine fauna in Western Australia. Detailed knowledge regarding all stages of a species lifecycle – such as migration patterns, spawning or calving areas, and foraging grounds – is rarely available. For a few species, such as those important for commercial and recreational fishers, scientific understanding is stronger and predictions can be made for a number of key parameters such as population sizes and recruitment. For other species, detailed knowledge around key life stages is available, such as the location and populations of turtles at nesting beaches and the locations of sea lion haul out areas, breeding islands and populations at those areas, but there may be very little information for the remaining life stages. For many other species, including some species of conservation listed marine fauna, little information is available regarding recruitment rates, population sizes, migration patterns or key foraging areas. Understanding at the marine and coastal ecosystem level is also highly variable, particularly for processes, functions and responses to pressures. In some areas, such as Ningaloo Reef and Cockburn Sound, significant research effort has been undertaken and these systems are relatively well understood. However there has not been a systematic survey across the majority of Western Australian's marine and coastal environment, and the scale, remoteness and difficulty of undertaking marine fauna surveys means that within every bioregion new species are still being found and ecosystem processes are still being investigated. This is particularly the case for remote areas of the State. For all proposals which have the potential to significantly impact on marine fauna the EPA will take into account the level of knowledge underpinning predictions of environmental impacts and risks. While region-wide studies are beyond the capacity of any individual proponent, the EPA strongly encourages the sharing of information on marine fauna so that there is a collective improvement and consolidation of knowledge over time. This increased certainty will result in more straight forward assessments and also lead to improved environmental outcomes.

Impacts

Development activities that have the potential to impact marine fauna include, but are not necessarily limited to:

- Activities that change the characteristics of the marine and coastal environment including:
    - Dredging and rock dumping through increases in turbidity and the mobilisation of contaminants located within the sediment.
    - Construction of breakwaters, bridges, rock walls and other marine infrastructure that has the potential to change marine currents and other coastal processes.

  - Water quality within confined water bodies, such as marinas and ports, and its exchange with the broader marine and coastal environment.
  - Outflows and discharges from commercial and industrial development and aquaculture operations.
  - Accidental discharges.
- Underwater noise including from pile driving, rock dumping, dredging, vessel movements and seismic exploration have the potential to generate underwater noise that can negatively impact marine fauna either through physical injury or avoidance behavior.
- Interaction with vessels and equipment, including from aquaculture developments, during construction and operation has the potential to injure or kill marine fauna or result in avoidance behavior.
- Inappropriate lighting on either the shore or on vessels that can alter natural turtle behavior during the nesting season.
- High volume seawater intakes (e.g. desalination and cooling water) can entrain or trap adult and juvenile marine fauna as well as large numbers of planktonic larvae and result in mortality.
- Introduction of invasive marine species through poor inspection of marine plant and equipment and through ballast water exchange.

For the purposes of EIA, the EPA is focussed on significant impacts to marine fauna, which can include:

- Harm of individuals and declines in the population or the range of species protected under state legislation.
- Reductions in populations of species of local and regional importance.
- Impacts to species or groups of species that fulfil critical ecological functions within the ecosystem.
- Loss or impact to critical marine fauna habitat, including habitats such as nesting beaches, nursery areas, sea lion haul out areas, specific foraging or breeding areas, and fish spawing aggregation areas.
- Reduction in species diversity in an area, which may be due to factors such as migration or range contraction resulting from a decline in the quality of the local environment.
- Introduction and spread of invasive marine species or diseases.

Phytoplankton

The word Phytoplankton is derived from the Greek words phyto (plant) and plankton (made to wander or drift), phytoplankton are microscopic organisms that live in watery environments, both salty and fresh.

Some phytoplankton are bacteria, some are protists, and most are single-celled plants. Among the common kinds are cyanobacteria, silica-encased diatoms, dinoflagellates, green algae, and chalk-coated coccolithophores.

Phytoplankton are extremely diverse, varying from photosynthesizing bacteria (cyanobacteria), to plant-like diatoms, to armor-plated coccolithophores (drawings not to scale).

Like land plants, phytoplankton have chlorophyll to capture sunlight, and they use photosynthesis to turn it into chemical energy. They consume carbon dioxide, and release oxygen. All phytoplankton photosynthesize, but some get additional energy by consuming other organisms.

Phytoplankton growth depends on the availability of carbon dioxide, sunlight, and nutrients. Phytoplankton, like land plants, require nutrients such as nitrate, phosphate, silicate, and calcium at various levels depending on the species. Some phytoplankton can fix nitrogen and can grow in areas where nitrate concentrations are low. They also require trace amounts of iron which limits phytoplankton growth in large areas of the ocean because iron concentrations are very low. Other factors influence phytoplankton growth rates, including water temperature and salinity, water depth, wind, and what kinds of predators are grazing on them.

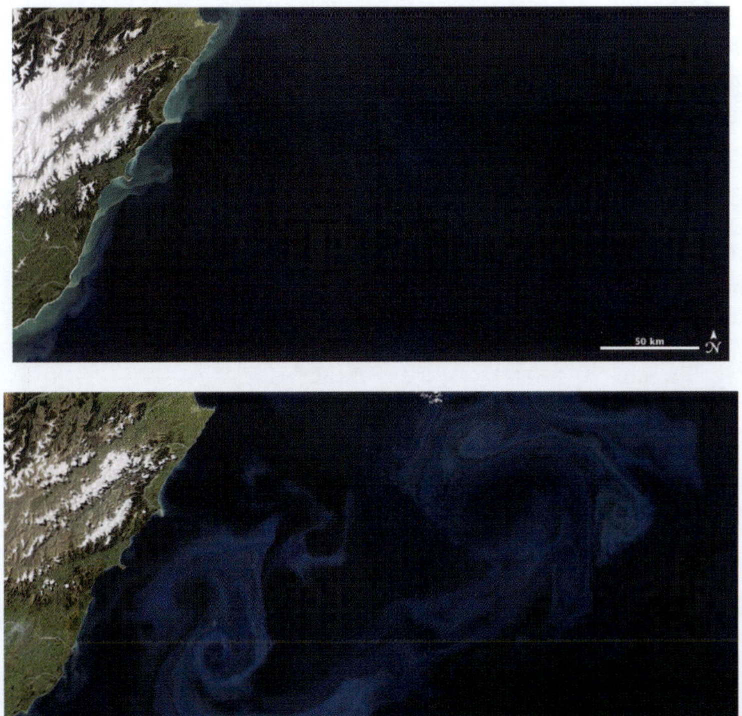

Phytoplankton can grow explosively over a few days or weeks. This pair of satellite images shows a bloom that formed east of New Zealand between October 11 and October 25, 2009.

When conditions are right, phytoplankton populations can grow explosively, a phenomenon known as a bloom. Blooms in the ocean may cover hundreds of square kilometers and are easily visible in satellite images. A bloom may last several weeks, but the life span of any individual phytoplankton is rarely more than a few days.

Importance of Phytoplankton

Food Web

Phytoplankton are the foundation of the aquatic food web, the primary producers, feeding everything from microscopic, animal-like zooplankton to multi-ton whales. Small fish and invertebrates also graze on the plant-like organisms, and then those smaller animals are eaten by bigger ones.

Phytoplankton can also be the harbingers of death or disease. Certain species of phytoplankton produce powerful biotoxins, making them responsible for so-called "red tides," or harmful algal blooms. These toxic blooms can kill marine life and people who eat contaminated seafood.

Dead fish washed onto a beach at Padre Island.

Phytoplankton cause mass mortality in other ways. In the aftermath of a massive bloom, dead phytoplankton sink to the ocean or lake floor. The bacteria that decompose the phytoplankton deplete the oxygen in the water, suffocating animal life; the result is a dead zone.

Climate and the Carbon Cycle

Through photosynthesis, phytoplankton consume carbon dioxide on a scale equivalent to forests and other land plants. Some of this carbon is carried to the deep ocean when phytoplankton die, and some is transferred to different layers of the ocean as phytoplankton are eaten by other creatures, which themselves reproduce, generate waste, and die.

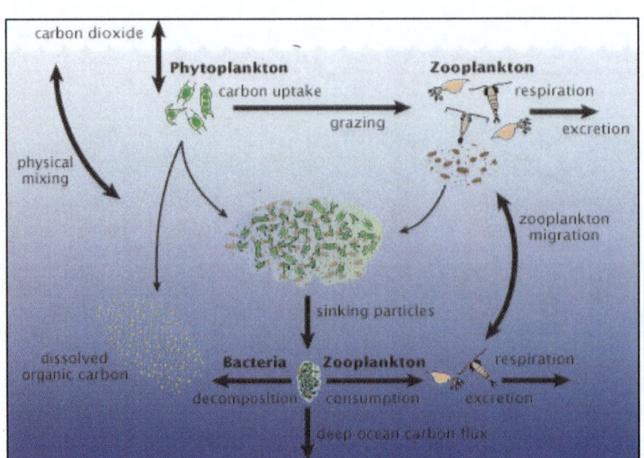

Phytoplankton are responsible for most of the transfer of carbon dioxide from the atmosphere to the ocean. Carbon dioxide is consumed during photosynthesis, and the carbon is incorporated in the phytoplankton, just as carbon is stored in the wood and leaves of a tree. Most of the carbon is returned to near-surface waters when phytoplankton are eaten or decompose, but some falls into the ocean depths.

Worldwide, this "biological carbon pump" transfers about 10 gigatonnes of carbon from the atmosphere to the deep ocean each year. Even small changes in the growth of phytoplankton may affect atmospheric carbon dioxide concentrations, which would feed back to global surface temperatures.

Aquatic Food Web

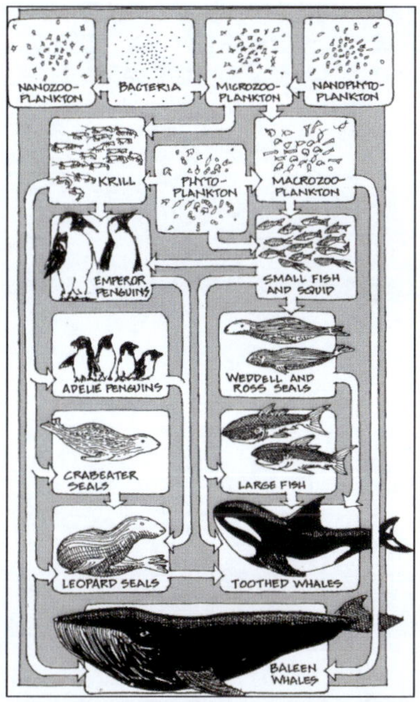

Phytoplankton form the base of the aquatic food web.

Studying Phytoplankton

Phytoplankton samples can be taken directly from the water at permanent observation stations or from ships. Sampling devices include hoses and flasks to collect water samples, and sometimes, plankton are collected on filters dragged through the water behind a ship.

Marine biologists use plankton nets to sample phytoplankton directly from the ocean.

Samples may be sealed and put on ice and transported for laboratory analysis, where researchers may be able to identify the phytoplankton collected down to the genus or even species level through microscopic investigation or genetic analysis.

Although samples taken from the ocean are necessary for some studies, satellites are pivotal for global-scale studies of phytoplankton and their role in climate change. Individual phytoplankton are tiny, but when they bloom by the billions, the high concentrations of chlorophyll and other light-catching pigments change the way the surface reflects light.

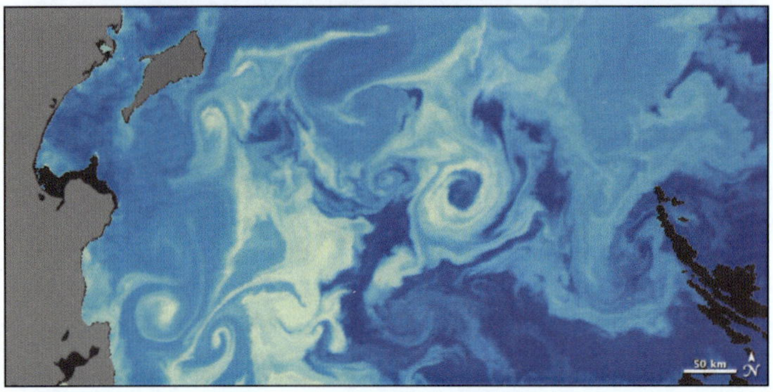

In natural-color satellite images (top), phytoplankton appear as colorful swirls. Scientists use these observations to estimate chlorophyll concentration (bottom) in the water.

The water may turn greenish, reddish, or brownish. The chalky scales that cover coccolithophores color the water milky white or bright blue. Scientists use these changes in ocean color to estimate chlorophyll concentration and the biomass of phytoplankton in the ocean.

Global Patterns and Cycles

Differences from Place to Place

Phytoplankton thrive along coastlines and continental shelves, along the equator in the Pacific and Atlantic Oceans, and in high-latitude areas. Winds play a strong role in the distribution of phytoplankton because they drive currents that cause deep water, loaded with nutrients, to be pulled up to the surface.

These upwelling zones, including one along the equator maintained by the convergence of the easterly trade winds, and others along the western coasts of several continents, are among the most productive ocean ecosystems. By contrast, phytoplankton are scarce in remote ocean gyres due to nutrient limitations.

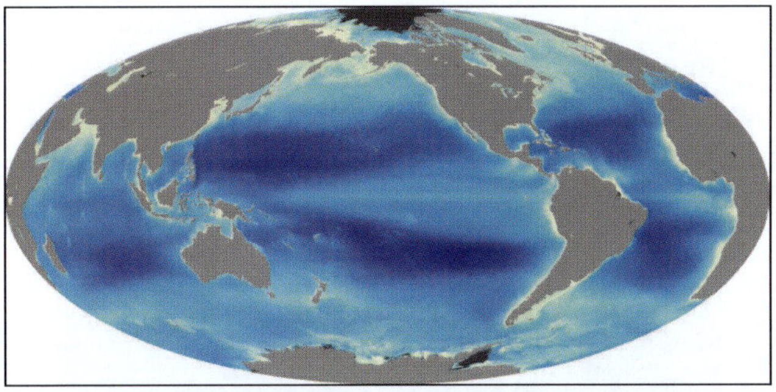

Phytoplankton are most abundant (yellow, high chlorophyll) in high latitudes and in upwelling zones along the equator and near coastlines. They are scarce in remote oceans (dark blue), where nutrient levels are low.

Differences from Season to Season

Like plants on land, phytoplankton growth varies seasonally. In high latitudes, blooms peak in the spring and summer, when sunlight increases and the relentless mixing of the water by winter storms subsides. Recent research suggests the vigorous winter mixing sets the stage for explosive spring growth by bringing nutrients up from deeper waters into the sunlit layers at the surface and separating phytoplankton from their zooplankton predators.

In the subtropical oceans, by contrast, phytoplankton populations drop off in summer. As surface waters warm up through the summer, they become very buoyant. With warm, buoyant water on top and cold, dense water below, the water column doesn't mix easily. Phytoplankton use up the nutrients available, and growth falls off until winter storms kick-start mixing.

In lower-latitude areas, including the Arabian Sea and the waters around Indonesia, seasonal blooms are often linked to monsoon-related changes in winds. As the winds reverse direction (offshore versus onshore), they alternately enhance or suppress upwelling, which changes nutrient concentrations. In the equatorial upwelling zone, there is very little seasonal change in phytoplankton productivity.

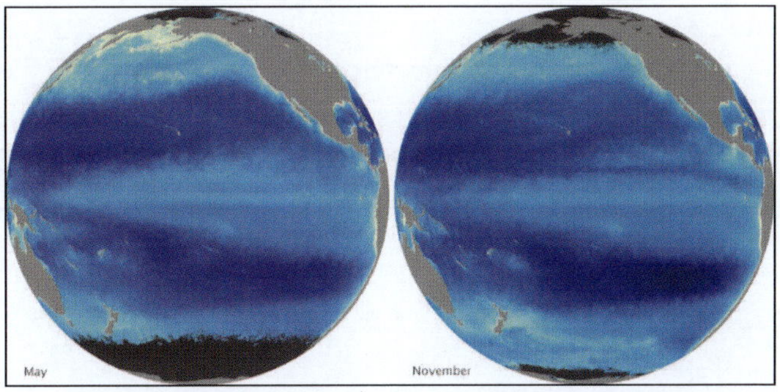

In spring and summer, phytoplankton bloom at high latitudes and decline in subtropical latitudes.

Differences from Year to Year

The biggest influence on year-to-year differences in global phytoplankton productivity is the El Niño-Southern Oscillation (ENSO) climate pattern. ENSO cycles are significant changes from typical sea surface temperatures, wind patterns, and rainfall in the Pacific Ocean along the equator.

During EL Niño events, phytoplankton productivity in the equatorial Pacific declines dramatically as the easterly trade winds that normally drive upwelling grow still or even reverse direction. The transition between El Niño and its counterpart, La Niña, is sometimes accompanied by a dramatic surge in phytoplankton productivity as upwelling of nutrient-rich deep water is suddenly renewed.

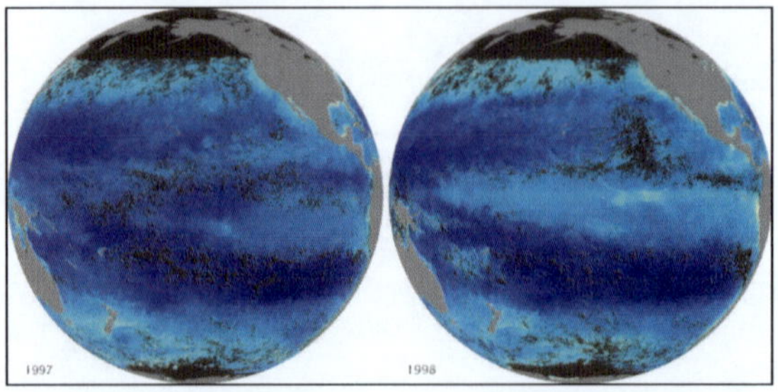

During an El Niño (December 1997, left), upwelling in the equatorial Pacific slows, reducing phytoplankton density. In contrast, a La Niña increases upwelling in the same area, enhancing phytoplankton growth (December 1998, right).

El Niño events influence weather patterns beyond the Pacific; in the eastern Indian Ocean around Indonesia, for example, phytoplankton productivity increases during El Niño. Productivity in the Gulf of Mexico and the western sub-tropical Atlantic has increased during El Niño events in the past decade, probably because increased rainfall and runoff delivered more nutrients than usual.

Compared to the ENSO-related changes in the productivity in the tropical Pacific, year-to-year differences in productivity in mid- and high latitudes are small.

Long-term Changes in Phytoplankton

Productivity

Because phytoplankton are so crucial to ocean biology and climate, any change in their productivity could have a significant influence on biodiversity, fisheries and the human food supply, and the pace of global warming.

Many models of ocean chemistry and biology predict that as the ocean surface warms in response to increasing atmospheric greenhouse gases, phytoplankton productivity will decline. Productivity is expected to drop because as the surface waters warm, the water column becomes increasingly stratified; there is less vertical mixing to recycle nutrients from deep waters back to the surface.

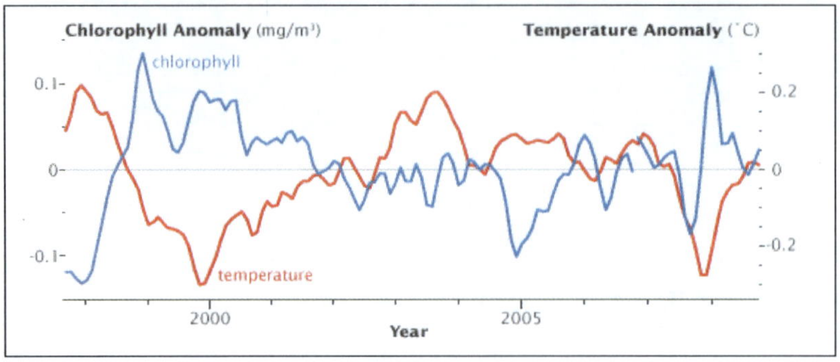

About 70% of the ocean is permanently stratified into layers that don't mix well. Between late 1997 and mid-2008, satellites observed that warmer-than-average temperatures (red line) led to below-average chlorophyll concentrations (blue line) in these areas.

Over the past decade, scientists have begun looking for this trend in satellite observations, and early studies suggest there has been a small decrease in global phytoplankton productivity. For example, ocean scientists documented an increase in the area of subtropical ocean gyres—the least productive ocean areas—over the past decade. These low-nutrient "marine deserts" appear to be expanding due to rising ocean surface temperatures.

Species Composition

Hundreds of thousands of species of phytoplankton live in Earth's oceans, each adapted to particular water conditions. Changes in water clarity, nutrient content, and salinity change the species that live in a given place.

Because larger plankton require more nutrients, they have a greater need for the vertical mixing of the water column that restocks depleted nutrients. As the ocean has warmed since the 1950s, it has become increasingly stratified, which cuts off nutrient recycling.

Continued warming due to the build up of carbon dioxide is predicted to reduce the amounts of larger phytoplankton such as diatoms, compared to smaller types, like cyanobacteria. Shifts in the relative abundance of larger versus smaller species of phytoplankton have been observed already in places around the world, but whether it will change overall productivity remains uncertain.

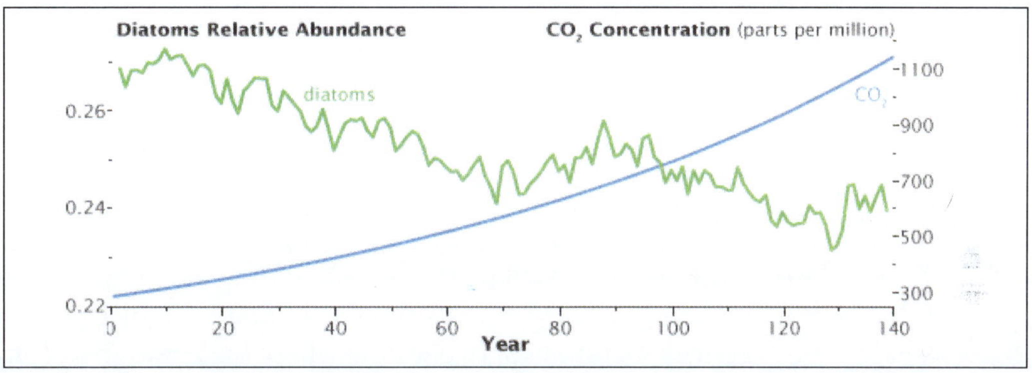

As carbon dioxide concentrations (blue line) increase in the next century, oceans will become more stratified. As upwelling declines, populations of larger phytoplankton such as diatoms are predicted to decline (green line).

These shifts in species composition may be benign, or they may result in a cascade of negative consequences throughout the marine food web. Accurate global mapping of phytoplankton taxonomic groups is one of the primary goals of proposed future NASA missions like the Aerosol, Cloud, Ecology (ACE) mission.

Zooplankton

Ecology

Zooplankton are heterotrophic (sometimes detritivorous) plankton (cf. phytoplankton). Plankton are organisms drifting in oceans, seas, and bodies of fresh water. The word zooplankton is derived from the Greek zoon, meaning "animal", and planktos, meaning "wanderer" or "drifter".

Individual zooplankton are usually microscopic, but some (such as jellyfish) are larger and visible to the naked eye.

A variety of zooplanktonic organisms.

Zooplankton is a categorization spanning a range of organism sizes including small protozoans and large metazoans. It includes holoplanktonic organisms whose complete life cycle lies within the plankton, as well as meroplanktonic organisms that spend part of their lives in the plankton before graduating to either the nekton or a sessile, benthic existence. Although zooplankton are primarily transported by ambient water currents, many have locomotion, used to avoid predators (as in diel vertical migration) or to increase prey encounter rate.

Ecologically important protozoan zooplankton groups include the foraminiferans, radiolarians and dinoflagellates (the last of these are often mixotrophic). Important metazoan zooplankton include cnidarians such as jellyfish and the Portuguese Man o' War; crustaceans such as copepods, ostracods, isopods, amphipods, mysids and krill; chaetognaths (arrow worms); molluscs such as pteropods; and chordates such as salps and juvenile fish. This wide phylogenetic range includes a similarly wide range in feeding behavior: filter feeding, predation and symbiosis with autotrophic phytoplankton as seen in corals. Zooplankton feed on bacterioplankton, phytoplankton, other zooplankton (sometimes cannibalistically), detritus (or marine snow) and even nektonic organisms. As a result, zooplankton are primarily found in surface waters where food resources (phytoplankton or other zooplankton) are abundant.

Segmented worm.

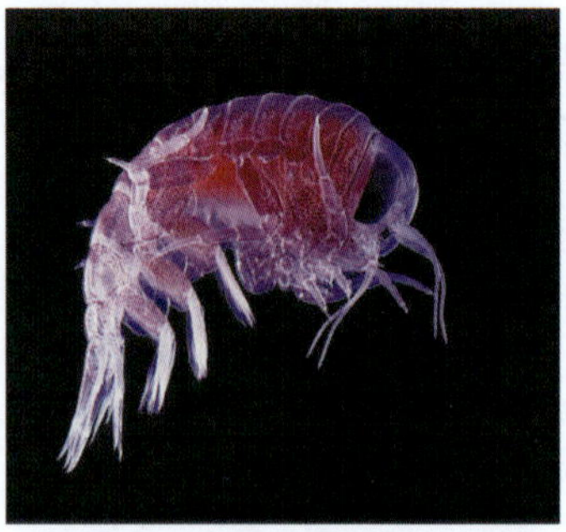

An amphipod.

Just as any species can be limited within a geographical region, so are zooplankton. However, species of zooplankton are not dispersed uniformly or randomly within a region of the ocean. As with phytoplankton, 'patches' of zooplankton species exist throughout the ocean. Though few physical barriers exist above the mesopelagic, specific species of zooplankton are strictly restricted by salinity and temperature gradients; while other species can withstand wide temperature and salinity gradients. Zooplankton patchiness can also be influenced by biological factors, as well as other physical factors. Biological factors include breeding, predation, concentration of phytoplankton, and vertical migration. The physical factor that influences zooplankton distribution the most is mixing of the water column (upwelling and downwelling along the coast and in the open ocean) that affects nutrient availability and, in turn, phytoplankton production.

Through their consumption and processing of phytoplankton and other food sources, zooplankton play a role in aquatic food webs, as a resource for consumers on higher trophic levels (including fish), and as a conduit for packaging the organic material in the biological pump. Since they are typically small, zooplankton can respond rapidly to increases in phytoplankton abundance, for instance, during the spring bloom.

Zooplankton can also act as a disease reservoir. Crustacean zooplankton have been found to house the bacterium Vibrio cholerae, which causes cholera, by allowing the cholera vibrios to attach to their chitinous exoskeletons. This symbiotic relationship enhances the bacterium's ability to survive in an aquatic environment, as the exoskeleton provides the bacterium with carbon and nitrogen.

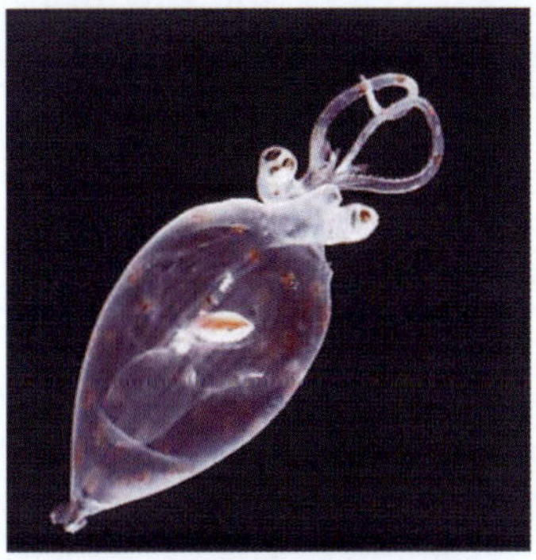

Juvenile planktonic squid.

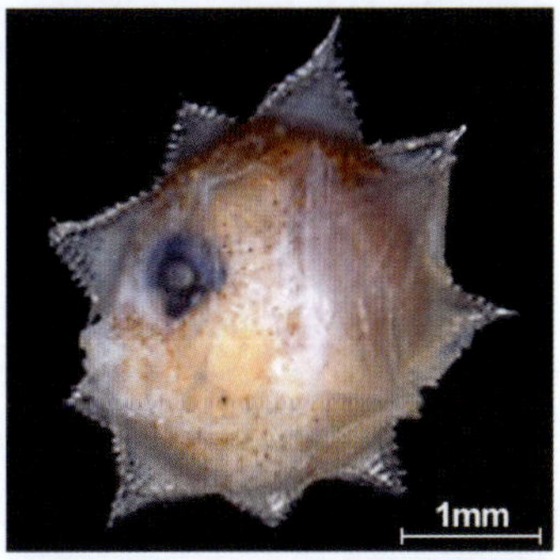

Ocean sunfish larvae (2.7mm).

Bacterioplankton

Bacterioplankton refers to the bacterial component of the plankton that drifts in the water column. They are found in both seawater and freshwater.

Bacterioplankton occupy a range of ecological niches in marine and aquatic ecosystems. They are both primary producers and primary consumers in these ecosystems and drive global biogeochemical cycling of elements essential for life (e.g., carbon and nitrogen). Many bacterioplankton species are autotrophic, and derive energy from either photosynthesis or chemosynthesis. Photosynthetic bacterioplankton are often categorized as picophytoplankton, and include major cyanobacterial groups such as Prochlorococcus and Synechococcus. Other heterotrophic bacterioplankton are saprotrophic, and obtain energy by consuming organic material produced by other organisms. This material may be dissolved in the medium and taken directly from there, or bacteria may live and grow in association with particulate material such as marine snow. Bacterioplankton play critical roles in global nitrogen fixation, nitrification, denitrification, remineralisation and methanogenesis.

Bacterioplankton abundance depends on environmental variables like temperature, nutrient availability and predation. Like other small plankton, the bacterioplankton are preyed upon by zooplankton (usually protozoans), and their numbers are also controlled through infection by bacteriophages.

Major Groups

Photosynthetic Bacterioplankton

Photosynthetic bacterioplankton are responsible for a large proportion of the total primary production of aquatic food webs, supplying organic compounds to higher trophic levels. These bacteria undergo oxygenic and anoxygenic photosynthesis. Differences between these processes can be seen in the byproducts produced, the primary electron donor, and the light harvesting pigments used for energy capture.

Cyanobacteria are a large group of photosynthetic bacterioplankton, often growing as cells or in filamentous colonies. These organisms are the dominant group of bacterioplankton using oxygenic photosynthesis in aquatic ecosystems. Cyanobacteria, along with photosynthetic eukaryotes, are responsible for approximately half of the total global primary production making them key players in the food web. They use photosynthesis to generate energy in the form of organic compounds and produce oxygen as a byproduct. Major light harvesting pigments include chlorophylls, phycoerytherin, phycocyanin and carotenoids. The majority of cyanobacteria found in marine environments are represented by the genera Synechococcus and Prochlorococcus. Synechococcus is cosmopolitan, having been reported across temperate and tropical waters. Prochlorococcus is a very small in size and is found mainly in the euphotic zone of tropical waters. Factors including light, nutrients, and temperature can cause cyanobacteria to proliferate and form harmful blooms. Cyanobacteria blooms can cause hypoxia and produce high levels of toxins, impacting other aquatic organisms as well as causing illnesses in humans.

Some Cyanobacteria are capable of nitrogen fixation. The genus Anabaena uses specialized cells called heterocysts to physically separate nitrogen fixation and photosynthesis. Trichodesmium is an example of cyanobacteria that is capable of fixing nitrogen through an alternative photosynthetic pathway.

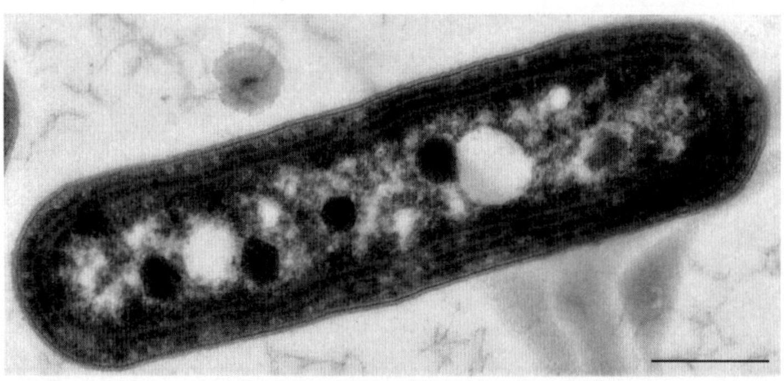

A transmission electron micrograph showing a species of the cyanobacteria Synechococcus.

Other photosynthetic bacterioplankton, including purple and green bacteria, undergo anoxygenic photosynthesis in anaerobic conditions. The pigments synthesized in these organisms are sensitive to oxygen. In purple bacteria the major pigments include bacteriochlorophyll a and b and carotenoids. Green bacteria have different light harvesting pigments consisting of bacteriochlorophyll c, d and e. These organisms do not produce oxygen through photosynthesis or use water as a reducing agent. Many of these organisms use sulfur, hydrogen or other compounds as an energy source to drive photosynthesis. Most of these bacterioplankton are found in anoxic waters, including stagnant and hypersaline environments.

Heterotrophic Bacterioplankton

Heterotrophic bacterioplankton rely on the available concentration of dissolved organic matter in the water column. Usually these organisms are saprophytic, absorbing nutrients from their surroundings. These heterotrophs also play a key role in the microbial loop and the remineralization of organic compounds like carbon and nitrogen. Pelagibacterales, also known as members of an alphaproteobacteria clade, are the most abundant bacterioplankton in the oceans. Members of this group are found in waters with low nutrient availability and are preyed on by protists.

Biogeochemical Cycling

Carbon

Atmospheric carbon is sequestered into the ocean by three main pumps which have been known for 30 years: the solubility pump, the carbonate pump, and the biological carbon pump (BCP). The biological carbon pump is a vertical transmission pump driven mainly by the sinking of organic rich particles. Bacterial phytoplankton near the surface incorporate atmospheric CO_2 and other nutrients into their biomass during photosynthesis. At the time of their death these phytoplankton, along with their incorporated carbon, sink to the bottom of the ocean where the carbon remains for thousands of years. The other biologically mediated sequestration of carbon in the ocean occurs through the microbial pump. The microbial pump is responsible for the production of old recalcitrant dissolved organic carbon (DOC) which is >100 years old. Plankton in the ocean are incapable of breaking down this recalcitrant DOC and thus it remains in the oceans for 1000's years without being respired. The two pumps work simultaneously, and the balance between them is believed to vary based on the availability of nutrients. Overall, the oceans act as a sink for atmospheric CO_2 but also release some carbon back into the atmosphere. This occurs when bacterioplankton and other organisms in the ocean consume organic matter and respire CO_2, and as a result of the solubility equilibrium between the ocean and the atmosphere.

Nitrogen

The nitrogen cycle in the oceans is mediated by microorganisms, many of which are bacteria, performing multiple conversions such as: nitrogen fixation, denitrification, assimilation, and anaerobic ammonia oxidation (anammox). There are many different nitrogen metabolism strategies employed by bacterioplankton. Starting with molecular nitrogen in the atmosphere (N_2), which is fixed by diazatrophs such as trichodesmium into usable forms like ammonia (NH_4). This ammonia can then be assimilated into organic matter like amino and nucleic acids, by both photoautrophic and heterotrophic plankton, it can also be nitrified to NO_3 for energy production by nitrifying bacteria. Finally the use of NO_3 or NO_2 as terminal electron acceptors reduces the nitrogen back into N_2, which is then released back into the atmosphere thus closing the cycle. Another important process involved in the regeneration of atmospheric N_2 is anammox. Anammox, a process in which ammonia is combined with nitrite in order to produce diatomic nitrogen and water, could account for 30-50% of production of N_2 in the ocean.

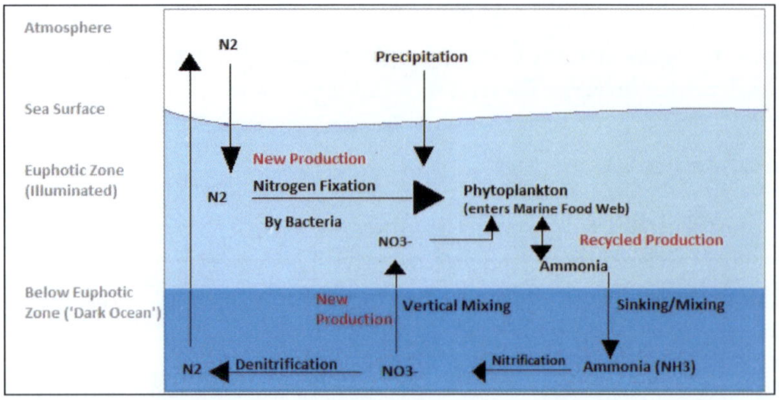

A schematic showing the cycling of nitrogen within the ocean.

Dissolved Organic Matter

Dissolved organic matter (DOM) is available in many forms in the ocean, and is responsible for supporting the growth of bacteria and microorganisms in the ocean. The two main sources of this dissolved organic matter are; decomposition of higher trophic level organisms like plants and fish, and secondly DOM in runoffs that pass through soil with high levels of organic material. It is important to note that the age and quality of the DOM is important for its usability by microbes. The majority of the DOM in the oceans is refractory or semi-labile and is not available for biodegradation. As mentioned above the microbial pump is responsible for the production of refractory DOM which is unavailable for biodegradation and remains dissolved in the oceans for thousands of years. The turnover of labile DOM organic material is quite high due to scarcity, this is important for the support of multiple trophic levels in the microbial community. The uptake and respiration of DOM by heterotrophs closes the cycle by producing CO_2.

Trophic Interactions

Variations in bacterioplankton abundance are usually a result of temperature, zooplankton grazing, and availability of substrate. Bacterial abundance and productivity are consistently related to algal abundance and productivity as well as organic carbon. Additionally, phosphorus directly influences both algal and bacterial abundance and in turn, algae and bacteria directly influence each other's abundance In extremely oligotrophic environments, both bacterial and algal growth is limited by phosphorus, but because bacteria are better competitors they obtain a larger portion of the inorganic substrate and increase in abundance more rapidly than algae.

In marine pelagic environments, heterotrophic nano-flagellates are the most probable consumers of bacterial cell production. Cultured flagellates in laboratory experiments demonstrate that they are adapted to predation on bacteria-sized particles and occur in concentrations to control bacterial biomass. Tight fluctuations in numbers of bacteria and flagellates have been found in a eutrophic estuary, particularly in the summer. The amplitude of these fluctuations increases in response to artificial eutrophication with inorganic nutrients and decreases in response to predation. Losses of bacterioplankton by grazing is indirectly related to carbon balances and directly related to prokaryotic inhibitors. A surplus of substrate would cause increased flagellate biomass, increased grazing on bacterioplankton and therefore decreased bacterial biomass overall. Predation of ciliates is analogous to predation by flagellates on bacteria as well.

With using prokaryotic inhibitors seasonally, there is a positive relationship between bacterial abundance and heterotrophic nanoplankton grazing rates and only 40-45 % of bacterioplankton production was observed to be consumed by phagotrophic Protozoa. Additionally, eukaryotic inhibitory experiments show that protozoan grazing has a positive effect on bacterioplankton production suggesting that nitrogen regeneration by Protozoa could be highly important for bacterial growth. Eukaryotic inhibitors did not prove to be useful to determine protozoan grazing rates on bacterioplankton, however they may help understand control mechanisms in the microbial food web.

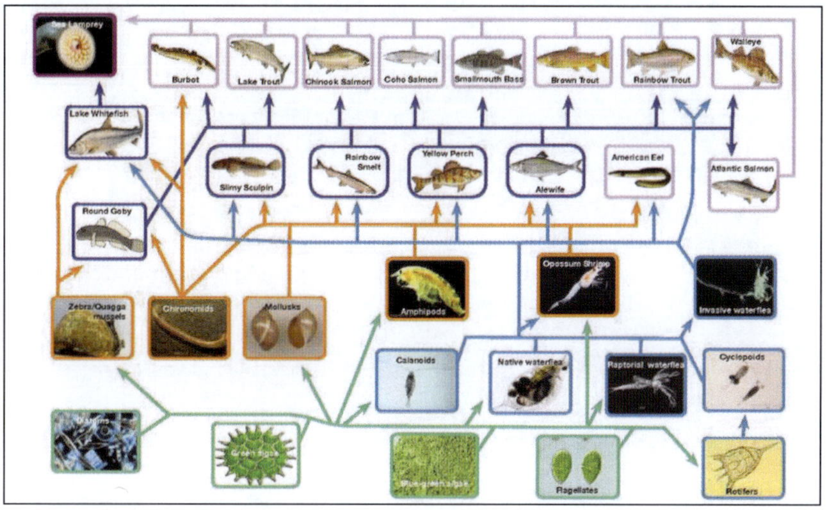

Schematic of the food chain in the freshwaters of Lake Ontario.

Ecological Significance

Bacterioplankton such as cyanobacteria are able to have toxic blooms in eutrophic lakes which can lead to the death of many organisms such as fish, birds, cattle, pets and humans. A few examples of these harmful blooms is the Microcystis bloom in the year 2000 in Swan River estuary, Australia, and the Oostvaarderplassen in the Netherlands in 2003. The detrimental effects of these blooms can range from heart malformation in fish to constraining copepod reproduction.

High temperatures caused by seasonality increases stratification and preventing vertical turbulent mixing which increases competition for light that favours buoyant cyanobacteria. Higher temperatures also reduce the viscosity of water which allows faster movement which also favors buoyant species of cyanobacteria. These species are also very competitive with the ability to create a surface cover preventing light to reach deeper species of plankton.

Climate studies are also indicating that with increasing hot waves the likelihood of detrimental cyanobacterial blooms will become more of a threat to eutrophic freshwater systems. Other implications of the increasing average air temperature due to climate change is that there might be an expansion of the cyanobacterial bloom season, extending from earlier in the spring to later in the fall. Estimates of bacterioplankton abundance and density can be derived with a variety of methods including direct counts, flow cytometry, and conclusions drawn from metabolic measures.

Further, as discussed in the biogeochemical cycling section, plankton are responsible for the recycling and movement of essential nutrients (i.e. nitrogen/carbon/DOM) which are essential building blocks for many of the organisms co-existing with bacterioplankton in these ecosystems. These recycled nutrients can be reused by primary producers, thus increasing the efficiency of the biological food web and minimizing energy waste.

A large harmful bloom of cyanobacteria, more commonly known as blue-green algae, spread across the lake in green filaments and strands that are clearly visible in this simulated-natural-colour image.

Marine Vertebrates

Vertebrates are a subphylum of chordates. They are chordates that have a vertebral column (backbone). The vertebral column provides the central support structure for an internal skeleton. The internal skeleton gives shape, support, and protection to the body and can provide a means of anchoring fins or limbs to the body. The vertebral column also serves to house and protect the spinal cord that lies within the column.

Marine vertebrates can be divided into marine fish and marine tetrapods.

Marine Fish

Fish fall into two main groups: fish with bony internal skeletons and fish with cartilaginous internal skeletons. Fish anatomy and physiology generally includes a two-chambered heart, eyes adapted to seeing underwater, and a skin protected by scales and mucous. They typically breathe by extracting oxygen from water through gills. Fish use fins to propel and stabilise themselves in the water. Over 33,000 species of fish have been described as of 2017, of which about 20,000 are marine fish.

Ray-finned fish.

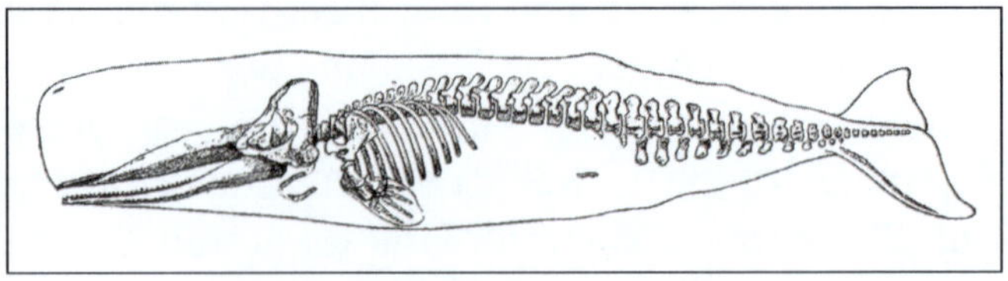

Marine tetrapod (sperm whale)
Skeletal structures showing the vertebral column running from the head to the tail.

Jawless Fish

Hagfish form a class of about 20 species of eel-shaped, slime-producing marine fish. They are the only known living animals that have a skull but no vertebral column. Lampreys form a superclass containing 38 known extant species of jawless fish. The adult lamprey is characterized by a toothed, funnel-like sucking mouth. Although they are well known for boring into the flesh of other fish to suck their blood, only 18 species of lampreys are actually parasitic. Together hagfish and lampreys are the sister group to vertebrates. Living hagfish remain similar to hagfish from around 300 million years ago. The lampreys are a very ancient lineage of vertebrates, though their exact relationship to hagfishes and jawed vertebrates is still a matter of dispute. Molecular analysis since 1992 has suggested that hagfish are most closely related to lampreys, and so also are vertebrates in a monophyletic sense. Others consider them a sister group of vertebrates in the common taxon of craniata.

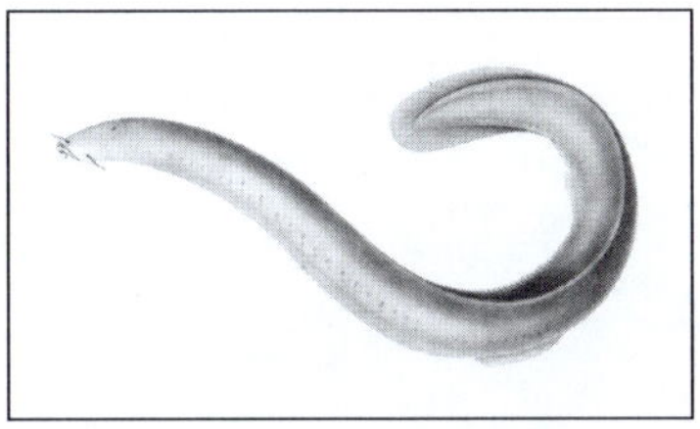

Hagfish are the only known living animals with a skull but no vertebral column.

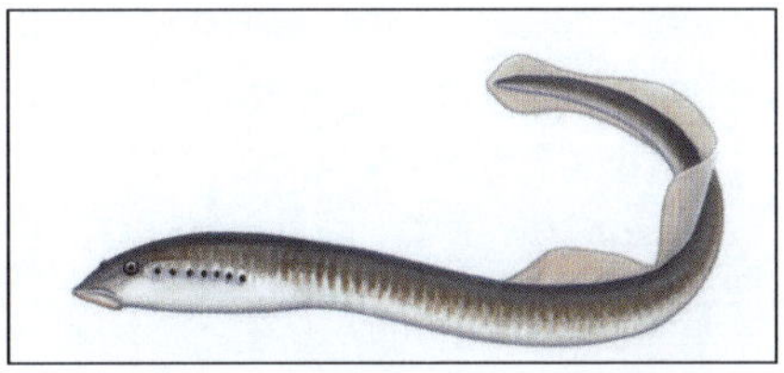

Lampreys are often parasitic and have a toothed, funnel-like sucking mouth.

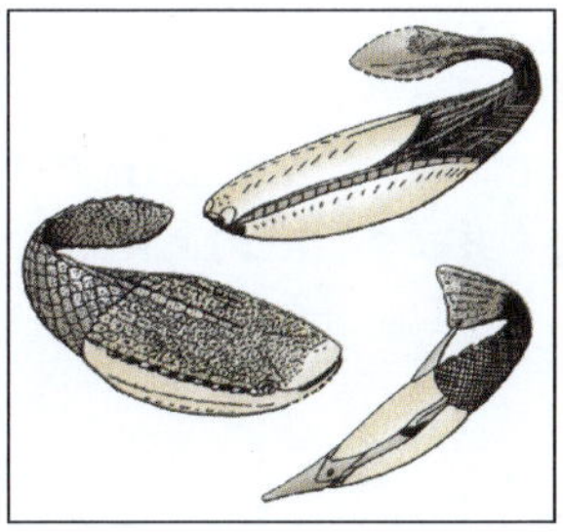

The extinct Pteraspidomorphi, ancestral to jawed vertebrates.

Pteraspidomorphi is an extinct class of early jawless fish ancestral to jawed vertebrates. The few characteristics they share with the latter are now considered as primitive for all vertebrates.

Cartilaginous Fish

Cartilaginous fish, such as sharks and rays, have jaws and skeletons made of cartilage rather than bone. Megalodon is an extinct species of shark that lived about 28 to 1.5 Ma. It looked much like a stocky version of the great white shark, but was much larger with fossil lengths reaching 20.3 metres (67 ft). Found in all oceans it was one of the largest and most powerful predators in vertebrate history, and probably had a profound impact on marine life. The Greenland shark has the longest known lifespan of all vertebrates, about 400 years.

Bony Fish

Bony fish have jaws and skeletons made of bone rather than cartilage. About 90% of the world's fish species are bony fish. Bony fish also have hard, bony plates called operculum which help them respire and protect their gills, and they often possess a swim bladder which they use for better control of their buoyancy.

Bony fish can be further divided into those with lobe fins and those with ray fins. Lobe fins have the form of fleshy lobes supported by bony stalks which extend from the body. Lobe fins evolved into the legs of the first tetrapod land vertebrates, so by extension an early ancestor of humans was a lobe-finned fish. Apart from the coelacanths and the lungfishes, lobe-finned fishes are now extinct. The rest of the modern fish have ray fins. These are made of webs of skin supported by bony or horny spines (rays) which can be erected to control the fin stiffness.

Marine Tetrapods

A tetrapod is a vertebrate with limbs (feet). Tetrapods evolved from ancient lobe-finned fishes about 400 million years ago during the Devonian Period when their earliest ancestors emerged from the sea and adapted to living on land. This change from a body plan for breathing and navigating in gravity-neutral water to a body plan with mechanisms enabling the animal to breath in air without dehydrating and move on land is one of the most profound evolutionary changes known. Tetrapods can be divided into four classes: amphibians, reptiles, birds and mammals.

Marine tetrapods are tetrapods that returned from land back to the sea again. The first returns to the ocean may have occurred as early as the Carboniferous Period whereas other returns occurred as recently as the Cenozoic, as in cetaceans, pinnipeds, and several modern amphibians.

Tiktaalik, an extinct lobe-finned fish, developed limb-like fins that could take it onto land.

Amphibians

Amphibians live part of their life in water and part on land. They mostly require fresh water to reproduce. A few inhabit brackish water, but there are no true marine amphibians. There have been reports, however, of amphibians invading marine waters, such as a Black Sea invasion by the natural hybrid Pelophylax esculentus reported in 2010.

Reptiles

Reptiles do not have an aquatic larval stage, and in this way are unlike amphibians. Most reptiles are oviparous, although several species of squamates are viviparous, as were some extinct aquatic clades—the fetus develops within the mother, contained in a placenta rather than an eggshell. As amniotes, reptile eggs are surrounded by membranes for protection and transport, which adapt them to reproduction on dry land. Many of the viviparous species feed their fetuses through various forms of placenta analogous to those of mammals, with some providing initial care for their hatchlings.

Some reptiles are more closely related to birds than other reptiles, and many scientists prefer to make Reptilia a monophyletic group which includes the birds. Extant non-avian reptiles which inhabit or frequent the sea include sea turtles, sea snakes, terrapins, the marine iguana, and the saltwater crocodile. Currently, of the approximately 12,000 extant reptile species and sub-species, only about 100 of are classed as marine reptiles.

Except for some sea snakes, most extant marine reptiles are oviparous and need to return to land to lay their eggs. Apart from sea turtles, the species usually spend most of their lives on or near land rather than in the ocean. Sea snakes generally prefer shallow waters nearby land, around islands, especially waters that are somewhat sheltered, as well as near estuaries. Unlike land snakes, sea snakes have evolved flattened tails which help them swim.

Marine iguana.

Leatherback sea turtle.

Saltwater crocodile.

Marine snakes have flattened tails.

The ancient Ichthyosaurus communis independently evolved flippers similar to dolphins.

Some extinct marine reptiles, such as ichthyosaurs, evolved to be viviparous and had no requirement to return to land. Ichthyosaurs resembled dolphins. They first appeared about 245 million years ago and disappeared about 90 million years ago. The terrestrial ancestor of the ichthyosaur had no features already on its back or tail that might have helped along the evolutionary process. Yet the ichthyosaur developed a dorsal and tail fin which improved its ability to swim. The biologist Stephen Jay Gould said the ichthyosaur was his favourite example of convergent evolution. The earliest marine reptiles arose in the Permian. During the Mesozoic many groups of reptiles became adapted to life in the seas, including ichthyosaurs, plesiosaurs, mosasaurs, nothosaurs, placodonts, sea turtles, thalattosaurs and thalattosuchians. Marine reptiles were less numerous after mass extinction at the end of the Cretaceous.

Birds

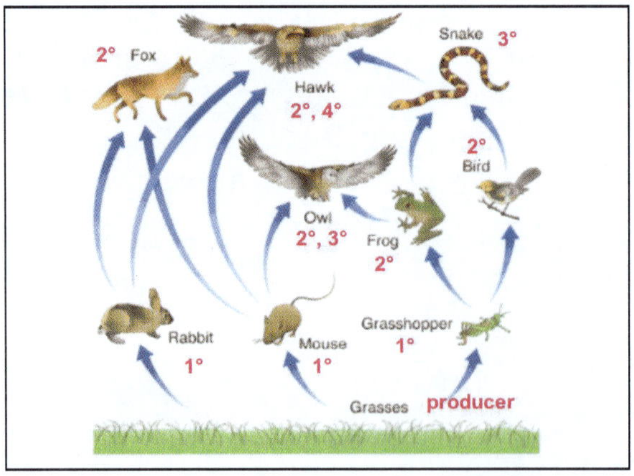

Waterbird food web in Chesapeake Bay.

Marine birds are adapted to life within the marine environment. They are often called seabirds. While marine birds vary greatly in lifestyle, behavior and physiology, they often exhibit striking convergent evolution, as the same environmental problems and feeding niches have resulted in similar adaptations. Examples include albatross, penguins, gannets, and auks.

In general, marine birds live longer, breed later and have fewer young than terrestrial birds do, but they invest a great deal of time in their young. Most species nest in colonies, which can vary in size from a few dozen birds to millions. Many species are famous for undertaking long annual migrations, crossing the equator or circumnavigating the Earth in some cases. They feed both at the ocean's surface and below it, and even feed on each other. Marine birds can be highly pelagic, coastal, or in some cases spend a part of the year away from the sea entirely. Some marine birds plummet from heights, plunging through the water leaving vapour-like trails, similar to that of fighter planes. Gannets plunge into the water at up to 100 kilometres per hour (60 mph). They have air sacs under their skin in their face and chest which act like bubble-wrap, cushioning the impact with the water.

Albatrosses range over huge areas of ocean and regularly circle the globe.

The first marine birds evolved in the Cretaceous period, and modern marine bird families emerged in the Paleogene.

Gannets "divebomb" at high speed.

Mammals

Mammals are characterised by the presence of mammary glands which in females produce milk for feeding (nursing) their young. There are about 130 living and recently extinct marine mammal species such as seals, dolphins, whales, manatees, sea otters and polar bears. They do not represent a distinct taxon or systematic grouping, but are instead unified by their reliance on the marine environment for feeding. Both cetaceans and sirenians are fully aquatic and therefore are obligate water dwellers. Seals and sea-lions are semiaquatic; they spend the majority of their time in the water, but need to return to land for important activities such as mating, breeding and molting. In contrast, both otters and the polar bear are much less adapted to aquatic living. Their diet varies considerably as well: some may eat zooplankton; others may eat fish, squid, shellfish, and sea-grass; and a few may eat other mammals.

In a process of convergent evolution, marine mammals such as dolphins and whales redeveloped their body plan to parallel the streamlined fusiform body plan of pelagic fish. Front legs became flippers and back legs disappeared, a dorsal fin reappeared and the tail morphed into a powerful horizontal fluke. This body plan is an adaptation to being an active predator in a high drag environment. A parallel convergence occurred with the now extinct ichthyosaur.

Sea otter, a classic keystone species which controls sea urchin numbers.

Endangered blue whale, largest animal ever.

Bottlenose dolphin, highest encephalization of any animal after humans.

Walrus.

Marine Reptile

Marine reptiles are reptiles which have become secondarily adapted for an aquatic or semiaquatic life in a marine environment.

The earliest marine reptiles arose in the Permian period during the Paleozoic era. During the Mesozoic era, many groups of reptiles became adapted to life in the seas, including such familiar clades as the ichthyosaurs, plesiosaurs (these two orders were once thought united in the group "Enaliosauria," a classification now cladistically obsolete), mosasaurs, nothosaurs, placodonts, sea turtles, thalattosaurs and thalattosuchians. After the mass extinction at the end of the Cretaceous period, marine reptiles were less numerous, but there was still a high variety of species in the early Cenozoic, such as "true" sea turtles, bothremydids, palaeophiid snakes, a few choristoderes such as Simoedosaurus and dyrosaurid crocodylomorphs. Various types of marine gavialid crocodilians remained widespread as recently as the Late Miocene.

Currently, of the approximately 12,000 extant reptile species and sub-species, only about 100 are classed as marine reptiles: extant marine reptiles include marine iguanas, sea snakes, sea turtles and saltwater crocodiles.

Some marine reptiles, such as ichthyosaurs, plesiosaurs, metriorhynchid thalattosuchians, and mosasaurs became so well adapted to a marine lifestyle that they were incapable of venturing onto land and gave birth in the water. Others, such as sea turtles and saltwater crocodiles, return to shore to lay their eggs. Some marine reptiles also occasionally rest.

Reptiles that live in the sea: Saltwater crocodile (top left), Sea turtle (top right), Marine iguana (bottom left), Sea snake (bottom right).

Extant (Living) Varieties

- Sea turtles: There are seven extant species of sea turtles, which live mostly along the tropical and subtropical coastlines, though some do migrate long distances and have been known to travel as far north as Scandinavia. Sea turtles are largely solitary animals, though some do form large, though often loosely connected groups during nesting season. Although only seven turtle species are truly marine, many more dwell in brackish waters.
- Sea snakes: The most abundant of the marine reptiles, there are over 60 different species of sea snakes. They inhabit the tropical and subtropical waters of the Indian and Pacific oceans, though very limited reports of sightings suggest they may be extending into the Atlantic ocean. Sea snakes are venomous and their bites have been known to be fatal, though generally they only bite when provoked and often inject only a very small, non-fatal quantity of venom. Sea snakes are distinguished from terrestrial snakes by a vertically flattened tail.

- Marine iguana: Marine iguanas live only on the Galápagos Islands and are not fully adapted to marine life. Although they feed exclusively on marine plants and spend a good deal of their time in the water, they do nest on land and need to bask in the sun to reach their ideal body temperature; they are thus also subject to terrestrial predators.

Saltwater and American crocodiles : none of the extant species of crocodiles is truly marine; however, the saltwater crocodile (Crocodylus porosus) does display adaptations to saltwater inhabitation and dwells in the brackish waters of Southeast Asia and Australia. Saltwater crocodiles dispose of excess salt in their bodies through specialized salt glands. These animals are the largest species of crocodiles, also making them the largest of the reptiles—they can grow up to six meters in length. American crocodiles (Crocodylus acutus) similarly prefer brackish over freshwater habitats.

Hawksbill sea turtle (Eretmochelys imbricata).

Most species of marine reptiles are considered endangered to some degree. All but one species of sea turtles are endangered due to destruction of nesting habitats on coastal lands, exploitation, and marine fishing; many species of sea snakes are threatened or endangered due to commercial exploitation (sale of skins) and pollution especially in Asia; marine iguanas are threatened due to their very limited habitation range. Saltwater crocodiles are at low risk for extinction.

Seabird

Seabirds (also known as marine birds) are birds that are adapted to life within the marine environment. While seabirds vary greatly in lifestyle, behavior and physiology, they often exhibit striking convergent evolution, as the same environmental problems and feeding niches have resulted in similar adaptations. The first seabirds evolved in the Cretaceous period, and modern seabird families emerged in the Paleogene.

In general, seabirds live longer, breed later and have fewer young than other birds do, but they invest a great deal of time in their young. Most species nest in colonies, which can vary in size from a few dozen birds to millions. Many species are famous for undertaking long annual migrations, crossing the equator or circumnavigating the Earth in some cases. They feed both at the ocean's surface and below it, and even feed on each other. Seabirds can be highly pelagic, coastal, or in some cases spend a part of the year away from the sea entirely.

Seabirds and humans have a long history together: they have provided food to hunters, guided fishermen to fishing stocks and led sailors to land. Many species are currently threatened by human activities, and conservation efforts are under way.

The sooty tern is highly aerial and marine and will spend months flying at sea, returning to land only for breeding.

Classification of Seabirds

There exists no single definition of which groups, families and species are seabirds, and most definitions are in some way arbitrary. In the words of two seabird scientists, "The one common characteristic that all seabirds share is that they feed in saltwater; but, as seems to be true with any statement in biology, some do not." However, by convention all of the Sphenisciformes and Procellariiformes, all of the Pelecaniformes except the darters, and some of the Charadriiformes (the skuas, gulls, terns, auks and skimmers) are classified as seabirds. The phalaropes are usually included as well, since although they are waders ("shorebirds" in North America), two of the three species are oceanic for nine months of the year, crossing the equator to feed pelagically.

Loons and grebes, which nest on lakes but winter at sea, are usually categorized as water birds, not seabirds. Although there are a number of sea ducks in the family Anatidae that are truly marine in the winter, by convention they are usually excluded from the seabird grouping. Many waders (or shorebirds) and herons are also highly marine, living on the sea's edge (coast), but are also not treated as seabirds. Sea eagles and other fish-eating birds of prey are also typically excluded, however tied to marine environments they may be.

Evolution and Fossil Record

Seabirds, by virtue of living in a geologically depositional environment (that is, in the sea where sediments are readily laid down), are well represented in the fossil record. They are first known to occur in the Cretaceous period, the earliest being the Hesperornithiformes, like Hesperornis regalis, a flightless loon-like seabird that could dive in a fashion similar to grebes and loons (using its feet to move underwater) but had a beak filled with sharp teeth. Flying Cretaceous seabirds do not exceed wingspans of two meters; any sizes were taken by piscivorous pterosaurs.

While Hesperornis is not thought to have left descendants, the earliest modern seabirds also occurred in the Cretaceous, with a species called Tytthostonyx glauconiticus, which seems allied to the Procellariiformes and Pelecaniformes. In the Paleogene both pterosaurs and marine reptiles became extinct, allowing seabirds to expand ecologically. These post-extinction seas were dominated by early Procellariidae, giant penguins and two extinct families, the Pelagornithidae and the Plotopteridae (a group of large seabirds that looked like the penguins). Modern genera began their wide radiation in the Miocene, although the genus Puffinus (which includes today's Manx shearwater and sooty shearwater) might date back to the Oligocene. The highest diversity of seabirds apparently existed during the Late Miocene and the Pliocene. At the end of the latter, the oceanic food web had undergone a period of upheaval due to extinction of considerable numbers of marine species; subsequently, the spread of marine mammals seems to have prevented seabirds from reaching their erstwhile diversity.

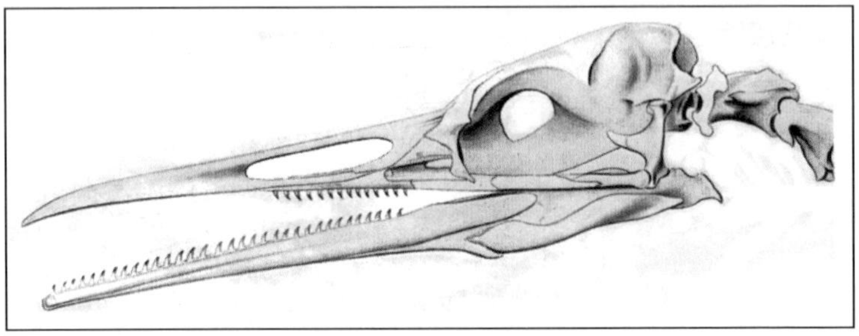

The Cretaceous seabird Hesperornis.

Characteristics

Adaptations to Life at Sea

Seabirds have made numerous adaptations to living on and feeding in the sea. Wing morphology has been shaped by the niche an individual species or family has evolved, so that looking at a wing's shape and loading can tell a scientist about its life feeding behavior. Longer wings and low wing loading are typical of more pelagic species, while diving species have shorter wings. Species such as the wandering albatross, which forage over huge areas of sea, have a reduced capacity for powered flight and are dependent on a type of gliding called dynamic soaring (where the wind deflected by waves provides lift) as well as slope soaring. Seabirds also almost always have webbed feet, to aid movement on the surface as well as assisting diving in some species. The Procellariiformes are unusual among birds in having a strong sense of smell, which is used to find widely distributed food in a vast ocean, and possibly to locate their colonies.

Salt glands are used by seabirds to deal with the salt they ingest by drinking and feeding (particularly on crustaceans), and to help them osmoregulate. The excretions from these glands (which are positioned in the head of the birds, emerging from the nasal cavity) are almost pure sodium chloride.

With the exception of the cormorants and some terns, and in common with most other birds, all seabirds have waterproof plumage. However, compared to land birds, they have far more feathers protecting their bodies. This dense plumage is better able to protect the bird from getting wet, and cold is kept out by a dense layer of down feathers. The cormorants possess a layer of unique feathers that retain a smaller layer of air (compared to other diving birds) but otherwise soak up water. This allows them to swim without fighting the buoyancy that retaining air in the feathers causes, yet retain enough air to prevent the bird losing excessive heat through contact with water.

The plumage of most seabirds is less colourful than that of land birds, restricted in the main to variations of black, white or grey. A few species sport colourful plumes (such as the tropicbirds and some penguins), but most of the colour in seabirds appears in the bills and legs. The plumage of seabirds is thought in many cases to be for camouflage, both defensive (the colour of US Navy battleships is the same as that of Antarctic prions, and in both cases it reduces visibility at sea) and aggressive (the white underside possessed by many seabirds helps hide them from prey below). The usually black wing tips help prevent wear, as they contain melanins to make them black that helps the feathers resist abrasion.

Cormorants, like this double-crested cormorant, have plumage that is partly wettable. This functional adaptation balances the competing requirement for thermoregulation against that of the need to reduce buoyancy.

Diet and Feeding

Seabirds evolved to exploit different food resources in the world's seas and oceans, and to a great extent, their physiology and behavior have been shaped by their diet. These evolutionary forces have often caused species in different families and even orders to evolve similar strategies and adaptations to the same problems, leading to remarkable convergent evolution, such as that between auks and penguins. There are four basic feeding strategies, or ecological guilds, for feeding at sea: surface feeding, pursuit diving, plunge diving and predation of higher vertebrates; within these guilds there are multiple variations on the theme.

Surface Feeding

Many seabirds feed on the ocean's surface, as the action of marine currents often concentrates food such as krill, forage fish, squid or other prey items within reach of a dipped head.

Surface feeding itself can be broken up into two different approaches, surface feeding while flying (for example as practiced by gadfly petrels, frigatebirds and storm petrels), and surface feeding while swimming (examples of which are practiced by fulmars, gulls, many of the shearwaters and gadfly petrels). Surface feeders in flight include some of the most acrobatic of seabirds, which either snatch morsels from the water (as do frigate-birds and some terns), or "walk", pattering and hovering on the water's surface, as some of the storm-petrels do. Many of these do not ever land in the water, and some, such as the frigatebirds, have difficulty getting airborne again should they do so. Another seabird family that does not land while feeding is the skimmer, which has a unique fishing method: flying along the surface with the lower mandible in the water—this shuts automatically when the bill touches something in the water. The skimmer's bill reflects its unusual lifestyle, with the lower mandible uniquely being longer than the upper one.

Surface feeders that swim often have unique bills as well, adapted for their specific prey. Prions have special bills with filters called lamellae to filter out plankton from mouthfuls of water, and many albatrosses and petrels have hooked bills to snatch fast-moving prey. Gulls have more generalised bills that reflect their more opportunistic lifestyle.

Wilson's storm petrels pattering on the water's surface.

Pursuit Diving

Pursuit diving exerts greater pressures (both evolutionary and physiological) on seabirds, but the reward is a greater area in which to feed than is available to surface feeders. Propulsion underwater can be provided by wings (as used by penguins, auks, diving petrels and some other species of petrel) or feet (as used by cormorants, grebes, loons and several types of fish-eating ducks). Wing-propelled divers are generally faster than foot-propelled divers. In both cases, the use of wings or feet for diving has limited their utility in other situations: loons and grebes walk with extreme difficulty (if at all), penguins cannot fly, and auks have sacrificed flight efficiency in favour of underwater diving. For example, the razorbill (an Atlantic auk) requires 64% more energy to fly than a petrel of equivalent size. Many shearwaters are intermediate between the two, having longer wings than typical wing-propelled divers but heavier wing loadings than the other surface-feeding procellariids, leaving them capable of diving to considerable depths while still being efficient long-distance travellers. The deepest diving exhibited by shearwaters is found in the short-tailed shearwater, which has been recorded diving below 70 m. Some albatross species are also capable of limited diving, with light-mantled sooty albatrosses holding the record at 12 m. Of all the wing-propelled pursuit divers, the most efficient in the air are the albatrosses, and they are also the poorest divers. This is the dominant guild in polar and subpolar environments, as it is energetically inefficient in warmer waters. With their poor flying ability, many wing-propelled pursuit divers are more limited in their foraging range than other guilds, especially during the breeding season when hungry chicks need regular feeding.

The chinstrap penguin is a highly streamlined pursuit diver.

Plunge Diving

Gannets, boobies, tropicbirds, some terns and brown pelicans all engage in plunge diving, taking fast moving prey by diving into the water from flight. Plunge diving allows birds to use the energy from the momentum of the dive to combat natural buoyancy (caused by air trapped in plumage), and thus uses less energy than the dedicated pursuit divers, allowing them to utilise more widely distributed food resources, for example, in impoverished tropical seas. In general, this is the most specialised method of hunting employed by seabirds; other non-specialists (such as gulls and skuas) may employ it but do so with less skill and from lower heights. In brown pelicans the skills of plunge diving take several years to fully develop—once mature, they can dive from 20 m (70 ft) above the water's surface, shifting the body before impact to avoid injury. It has been suggested that plunge divers are restricted in their hunting grounds to clear waters that afford a view of their prey from the air, and while they are the dominant guild in the tropics, the link between plunge diving and water clarity is inconclusive. Some plunge divers (as well as some surface feeders) are dependent on dolphins and tuna to push shoaling fish up towards the surface.

Kleptoparasitism, Scavenging and Predation

This catch-all category refers to other seabird strategies that involve the next trophic level up. Kleptoparasites are seabirds that make a part of their living stealing food of other seabirds. Most famously, frigatebirds and skuas engage in this behavior, although gulls, terns and other species will steal food opportunistically. The nocturnal nesting behavior of some seabirds has been interpreted as arising due to pressure from this aerial piracy. Kleptoparasitism is not thought to play a significant part of the diet of any species, and is instead a supplement to food obtained by hunting. A study of great frigatebirds stealing from masked boobies estimated that the frigatebirds could at most obtain 40% of the food they needed, and on average obtained only 5%. Many species of gull will feed on seabird and sea mammal carrion when the opportunity arises, as will giant petrels. Some species of albatross also engage in scavenging: an analysis of regurgitated squid beaks has shown that many of the squid eaten are too large to have been caught alive, and include mid-water species likely to be beyond the reach of albatrosses. Some species will also feed on other seabirds; for example, gulls, skuas and pelicans will often take eggs, chicks and even small adult seabirds from nesting colonies, while the giant petrels can kill prey up to the size of small penguins and seal pups.

Seabirds' life histories are dramatically different from those of land birds. In general, they are K-selected, live much longer (anywhere between twenty and sixty years), delay breeding for longer (for up to ten years), and invest more effort into fewer young. Most species will only have one clutch a year, unless they lose the first (with a few exceptions, like the Cassin's auklet), and many species (like the tubenoses and sulids), only one egg a year.

Care of young is protracted, extending for as long as six months, among the longest for birds. For example, once common guillemot chicks fledge, they remain with the male parent for several months at sea. The frigatebirds have the longest period of parental care of any bird except a few raptors and the southern ground hornbill, with each chick fledging after four to six months and continued assistance after that for up to fourteen months. Due to the extended period of care, breeding occurs every two years rather than annually for some species. This life-history strategy has probably evolved both in response to the challenges of living at sea (collecting widely scattered prey items), the frequency of breeding failures due to unfavourable marine conditions, and the relative lack of predation compared to that of land-living birds.

Because of the greater investment in raising the young and because foraging for food may occur far from the nest site, in all seabird species except the phalaropes, both parents participate in caring for the young, and pairs are typically at least seasonally monogamous. Many species, such as gulls, auks and penguins, retain the same mate for several seasons, and many petrel species mate for life. Albatrosses and procellariids, which mate for life, take many years to form a pair bond before they breed, and the albatrosses have an elaborate breeding dance that is part of pair-bond formation.

Northern gannet pair "billing" during courtship; like all seabirds except the phalaropes they maintain a pair bond throughout the breeding season.

Breeding and Colonies

Ninety-five percent of seabirds are colonial, and seabird colonies are among the largest bird colonies in the world, providing one of Earth's great wildlife spectacles. Colonies of over a million birds have been recorded, both in the tropics (such as Kiritimati in the Pacific) and in the polar latitudes (as in Antarctica). Seabird colonies occur exclusively for the purpose of breeding; non-breeding birds will only collect together outside the breeding season in areas where prey species are densely aggregated.

Seabird colonies are highly variable. Individual nesting sites can be widely spaced, as in an albatross colony, or densely packed as with a murre colony. In most seabird colonies, several different species will nest on the same colony, often exhibiting some niche separation. Seabirds can nest in trees (if any are available), on the ground (with or without nests), on cliffs, in burrows under the ground and in rocky crevices. Competition can be strong both within species and between species, with aggressive species such as sooty terns pushing less dominant species out of the most desirable nesting spaces. The tropical Bonin petrel nests during the winter to avoid competition with the more aggressive wedge-tailed shearwater. When the seasons overlap, the wedge-tailed shearwaters will kill young Bonin petrels in order to use their burrows.

Many seabirds show remarkable site fidelity, returning to the same burrow, nest or site for many years, and they will defend that site from rivals with great vigour. This increases breeding success, provides a place for returning mates to reunite, and reduces the costs of prospecting for a new site. Young adults breeding for the first time usually return to their natal colony, and often nest close to where they hatched. This tendency, known as philopatry, is so strong that a study of Laysan albatrosses found that the average distance between hatching site and the site where a bird established its own territory was 22 m; another study, this time on Cory's shearwaters nesting near Corsica, found that of nine out of 61 male chicks that returned to breed at their natal colony bred in the burrow they were raised in, and two actually bred with their own mother.

Colonies are usually situated on islands, cliffs or headlands, which land mammals have difficulty accessing. This is thought to provide protection to seabirds, which are often very clumsy on land. Coloniality often arises in types of bird that do not defend feeding territories (such as swifts, which have a very variable prey source); this may be a reason why it arises more frequently in seabirds. There are other possible advantages: colonies may act as information centres, where seabirds returning to the sea to forage can find out where prey is by studying returning individuals of the same species. There are disadvantages to colonial life, particularly the spread of disease. Colonies also attract the attention of predators, principally other birds, and many species attend their colonies nocturnally to avoid predation.

Common murres breed on densely packed colonies on offshore rocks, islands and cliffs.

Migration

Like many birds, seabirds often migrate after the breeding season. Of these, the trip taken by the Arctic tern is the farthest of any bird, crossing the equator in order to spend the Austral summer in Antarctica. Other species also undertake trans-equatorial trips, both from the north to the south, and from south to north. The population of elegant terns, which nest off Baja California, splits after the breeding season with some birds travelling north to the Central Coast of California and some travelling as far south as Peru and Chile to feed in the Humboldt Current. The sooty shearwater undertakes an annual migration cycle that rivals that of the Arctic tern; birds that nest in New Zealand and Chile and spend the northern summer feeding in the North Pacific off Japan, Alaska and California, an annual round trip of 40,000 statute miles (64,000 km).

Other species also migrate shorter distances away from the breeding sites, their distribution at sea determined by the availability of food. If oceanic conditions are unsuitable, seabirds will emigrate to more productive areas, sometimes permanently if the bird is young. After fledging, juvenile birds often disperse further than adults, and to different areas, so are commonly sighted far from a species' normal range. Some species, such as the auks, do not have a concerted migration effort, but drift southwards as the winter approaches. Other species, such as some of the storm petrels, diving petrels and cormorants, never disperse at all, staying near their breeding colonies year round.

Pelican flock flying over Havana Bay area. These birds come to Cuba every year from North America in the northern hemisphere winter season.

Away from the Sea

While the definition of seabirds suggests that the birds in question spend their lives on the ocean, many seabird families have many species that spend some or even most of their lives inland away from the sea. Most strikingly, many species breed tens, hundreds or even thousands of miles inland. Some of these species still return to the ocean to feed; for example, the snow petrel, the nests of which have been found 480 kilometres (300 mi) inland on the Antarctic mainland, are unlikely to find anything to eat around their breeding sites. The marbled murrelet nests inland in old growth forest, seeking huge conifers with large branches to nest on. Other species, such as the California gull, nest and feed inland on lakes, and then move to the coasts in the winter. Some cormorant, pelican, gull and tern species have individuals that never visit the sea at all, spending their lives on lakes, rivers, swamps and, in the case of some of the gulls, cities and agricultural land. In these cases it is thought that these terrestrial or freshwater birds evolved from marine ancestors. Some seabirds, principally those that nest in tundra, as skuas and phalaropes do, will migrate over land as well.

The more marine species, such as petrels, auks and gannets, are more restricted in their habits, but are occasionally seen inland as vagrants. This most commonly happens to young inexperienced birds, but can happen in great numbers to exhausted adults after large storms, an event known as a wreck, where they provide prized sightings for birders.

Arctic terns breed in the arctic and subarctic and winter in Antarctica.

Relationship with Humans

Seabirds and Fisheries

Seabirds have had a long association with both fisheries and sailors, and both have drawn benefits and disadvantages from the relationship.

Fishermen have traditionally used seabirds as indicators of both fish shoals, underwater banks that might indicate fish stocks, and of potential landfall. In fact, the known association of seabirds with land was instrumental in allowing the Polynesians to locate tiny landmasses in the Pacific. Seabirds have provided food for fishermen away from home, as well as bait. Famously, tethered cormorants have been used to catch fish directly. Indirectly, fisheries have also benefited from guano from colonies of seabirds acting as fertilizer for the surrounding seas.

Negative effects on fisheries are mostly restricted to raiding by birds on aquaculture, although long-lining fisheries also have to deal with bait stealing. There have been claims of prey depletion by seabirds of fishery stocks, and while there is some evidence of this, the effects of seabirds are considered smaller than that of marine mammals and predatory fish (like tuna).

Some seabird species have benefited from fisheries, particularly from discarded fish and offal. These discards compose 30% of the food of seabirds in the North Sea, for example, and compose up to 70% of the total food of some seabird populations. This can have other impacts; for example, the spread of the northern fulmar through the United Kingdom is attributed in part to the availability of discards. Discards generally benefit surface feeders, such as gannets and petrels, to the detriment of pursuit divers like penguins.

Fisheries also have negative effects on seabirds, and these effects, particularly on the long-lived and slow-breeding albatrosses, are a source of increasing concern to conservationists. The bycatch of seabirds entangled in nets or hooked on fishing lines has had a big impact on seabird numbers; for example, an estimated 100,000 albatrosses are hooked and drown each year on tuna lines set out by long-line fisheries. Overall, many hundreds of thousands of birds are trapped and killed each year, a source of concern for some of the rarest species (for example, only about 2,000 short-tailed albatrosses are known to still exist). Seabirds are also thought to suffer when overfishing occurs.

Seabirds (mostly northern fulmars) flocking at a long-lining vessel.

Exploitation

The hunting of seabirds and the collecting of seabird eggs have contributed to the declines of many species, and the extinction of several, including the great auk and the spectacled cormorant. Seabirds have been hunted for food by coastal peoples throughout history—one of the earliest instances known is in southern Chile, where archaeological excavations in middens has shown hunting of albatrosses, cormorants and shearwaters from 5000 BP. This pressure has led to some species becoming extinct in many places; in particular, at least 20 species of an original 29 no longer breed on Easter Island. In the 19th century, the hunting of seabirds for fat deposits and feathers for the millinery trade reached industrial levels. Muttonbirding (harvesting shearwater chicks) developed as important industries in both New Zealand and Tasmania, and the name of one species, the providence petrel, is derived from its seemingly miraculous arrival on Norfolk Island where it provided a windfall for starving European settlers. In the Falkland Islands, hundreds of thousands of penguins were harvested for their oil each year. Seabird eggs have also long been an important source of food for sailors undertaking long sea voyages, as well as being taken when settlements grow in areas near a colony. Eggers from San Francisco took almost half a million eggs a year from the Farallon Islands in the mid-19th century, a period in the islands' history from which the seabird species are still recovering.

Both hunting and egging continue today, although not at the levels that occurred in the past, and generally in a more controlled manner. For example, the Māori of Stewart Island/Rakiura continue to harvest the chicks of the sooty shearwater as they have done for centuries, using traditional methods (called kaitiakitanga) to manage the harvest, but now work with the University of Otago in studying the populations. In Greenland, however, uncontrolled hunting is pushing many species into steep decline.

Other Threats

Other human factors have led to declines and even extinctions in seabird populations, colonies and species. Of these, perhaps the most serious are introduced species. Seabirds, breeding predominantly on small isolated islands, have lost many predator defence behaviors. Feral cats are capable of taking seabirds as large as albatrosses, and many introduced rodents, such as the Pacific rat, can take eggs hidden in burrows. Introduced goats, cattle, rabbits and other herbivores can lead to problems, particularly when species need vegetation to protect or shade their young. Disturbance of breeding colonies by humans is often a problem as well—visitors, even well-meaning tourists, can flush brooding adults off a colony leaving chicks and eggs vulnerable to predators.

The build-up of toxins and pollutants in seabirds is also a concern. Seabirds, being apex predators, suffered from the ravages of DDT until it was banned; among other effects, DDT was implicated in embryo development problems and the skewed sex ratio of western gulls in southern California. Oil spills are also a threat to seabird species, as both a toxin and because the feathers of the birds become saturated by the oil, causing them to lose their waterproofing. Oil pollution threatens species with restricted ranges or already depressed populations.

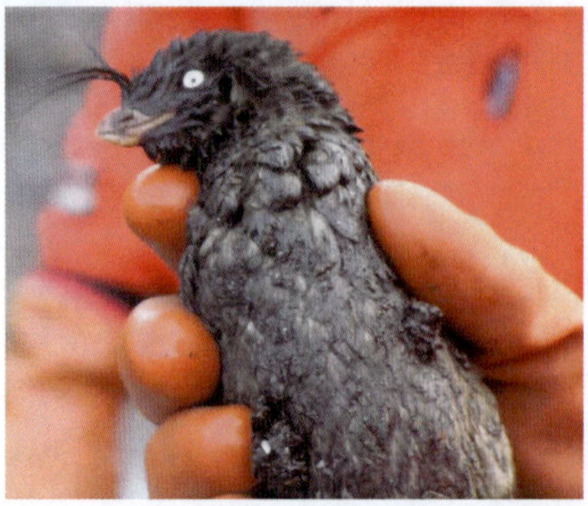

This crested auklet was oiled in Alaska during the M/V Selendang Ayu spill of 2004.

The threats faced by seabirds have not gone unnoticed by scientists or the conservation movement. As early as 1903, U.S. President Theodore Roosevelt was convinced of the need to declare Pelican

Island in Florida a National Wildlife Refuge to protect the bird colonies (including the nesting brown pelicans), and in 1909 he protected the Farallon Islands. Today many important seabird colonies are given some measure of protection, from Heron Island in Australia to Triangle Island in British Columbia.

Island restoration techniques, pioneered by New Zealand, enable the removal of exotic invaders from increasingly large islands. Feral cats have been removed from Ascension Island, Arctic foxes from many islands in the Aleutian Islands, and rats from Campbell Island. The removal of these introduced species has led to increases in numbers of species under pressure and even the return of extirpated ones. After the removal of cats from Ascension Island, seabirds began to nest there again for the first time in over a hundred years.

Seabird mortality caused by long-line fisheries can be greatly reduced by techniques such as setting long-line bait at night, dying the bait blue, setting the bait underwater, increasing the amount of weight on lines and by using bird scarers, and their deployment is increasingly required by many national fishing fleets. The international ban on the use of drift nets has also helped reduce the mortality of seabirds and other marine wildlife.

One of the Millennium Projects in the UK was the Scottish Seabird Centre, near the important bird sanctuaries on Bass Rock, Fidra and the surrounding islands. The area is home to huge colonies of gannets, puffins, skuas and other seabirds. The centre allows visitors to watch live video from the islands as well as learn about the threats the birds face and how we can protect them, and has helped to significantly raise the profile of seabird conservation in the UK. Seabird tourism can provide income for coastal communities as well as raise the profile of seabird conservation. For example, the northern royal albatross colony at Taiaroa Head in New Zealand attracts 40,000 visitors a year.

The plight of albatross and large seabirds, as well as other marine creatures, being taken as bycatch by long-line fisheries, has been addressed by a large number of non-governmental organizations (including BirdLife International, the American Bird Conservancy and the Royal Society for the Protection of Birds). This led to the Agreement on the Conservation of Albatrosses and Petrels, a legally binding treaty designed to protect these threatened species, which has been ratified by eleven countries as of 2008 (namely Argentina, Australia, Chile, Ecuador, France, New Zealand, Norway, Peru, South Africa, Spain and the United Kingdom).

Role in Culture

Many seabirds are little studied and poorly known, due to living far out to sea and breeding in isolated colonies. However, some seabirds, particularly, the albatrosses and gulls, have broken into popular consciousness. The albatrosses have been described as "the most legendary of birds", and have a variety of myths and legends associated with them, and today it is widely considered unlucky to harm them, although the notion that sailors believed that is a myth that derives from Samuel Taylor Coleridge's famous poem, "The Rime of the Ancient Mariner", in which a sailor is punished for killing an albatross by having to wear its corpse around his neck.

Sailors did, however, consider it unlucky to touch a storm petrel, especially one that has landed on the ship.

Gulls are one of the most commonly seen seabirds, given their use of human-made habitats (such as cities and dumps) and their often fearless nature. They therefore also have made it into the popular consciousness – they have been used metaphorically, as in Jonathan Livingston Seagull by Richard Bach, or to denote a closeness to the sea, such as their use in The Lord of the Rings – both in the insignia of Gondor and therefore Númenor (used in the design of the films), and to call Legolas to (and across) the sea. Other species have also made an impact; pelicans have long been associated with mercy and altruism because of an early Western Christian myth that they split open their breast to feed their starving chicks.

Depiction of a pelican with chicks on a stained glass window, Saint Mark's Church, Gillingham, Kent.

Seabird Families

The following are the groups of birds normally classed as seabirds.

- Sphenisciformes (Antarctic and southern waters; 16 species):
    - Spheniscidae penguins.
- G Procellariiformes (Tubenoses: pan-oceanic and pelagic; 93 species):
    - Diomedeidae albatrosses.
    - Procellariidae fulmars, prions, shearwaters, gadfly and other petrels.
    - Pelacanoididae diving petrels.
    - Hydrobatidae storm petrels.
- Pelecaniformes (Worldwide; 8 species) :
    - Pelecanidae pelicans.
- Suliformes (Worldwide; about 56 species):
    - Sulidae gannets and boobies.
    - Phalacrocoracidae cormorants.
    - Fregatidae frigatebirds.

- Phaethontiformes (Worldwide tropical seas; 3 species):
    - Phaethontidae tropicbirds.
- Charadriiformes (Worldwide; 305 species, but only the families listed are classed as seabirds):
    - Stercorariidae skuas.
    - Laridae gulls.
    - Sternidae terns.
    - Rhynchopidae skimmers.
    - Alcidae auks.

Marine Mammal

Marine mammals are aquatic mammals that rely on the ocean and other marine ecosystems for their existence. They include animals such as seals, whales, manatees, sea otters and polar bears. They do not represent a distinct taxon or systematic grouping, but rather have a polyphyletic relation due to convergent evolution, as in they do not have an immediate common ancestor. They are also unified by their reliance on the marine environment for feeding.

Marine mammal adaptation to an aquatic lifestyle varies considerably between species. Both cetaceans and sirenians are fully aquatic and therefore are obligate water dwellers. Seals and sea-lions are semiaquatic; they spend the majority of their time in the water but need to return to land for important activities such as mating, breeding and molting. In contrast, both otters and the polar bear are much less adapted to aquatic living. Their diet varies considerably as well; some may eat zooplankton, others may eat fish, squid, shellfish, sea-grass and a few may eat other mammals. While the number of marine mammals is small compared to those found on land, their roles in various ecosystems are large, especially concerning the maintenance of marine ecosystems, through processes including the regulation of prey populations. This role in maintaining ecosystems makes them of particular concern as 23% of marine mammal species are currently threatened.

Marine mammals were first hunted by aboriginal peoples for food and other resources. Many were also the target for commercial industry, leading to a sharp decline in all populations of exploited species, such as whales and seals. Commercial hunting led to the extinction of Steller's sea cow, sea mink, Japanese sea lion and the Caribbean monk seal. After commercial hunting ended, some species, such as the gray whale and northern elephant seal, have rebounded in numbers; conversely, other species, such as the North Atlantic right whale, are critically endangered. Other than hunting, marine mammals can be killed as bycatch from fisheries, where they become entangled in fixed netting and drown or starve. Increased ocean traffic causes collisions between fast ocean vessels and large marine mammals. Habitat degradation also threatens marine mammals and their ability to find and catch food. Noise pollution, for example, may adversely affect echolocating mammals, and the ongoing effects of global warming degrade Arctic environments.

Marine mammals form a diverse group of 129 species that rely on the ocean for their existence. They do not represent a distinct taxon or systematic grouping, but instead have a polyphyletic relationship. They are also unified by their reliance on the marine environment for feeding. Despite the diversity in morphology seen between groups, improved foraging efficiency has been the main driver in their evolution. The level of dependence on the marine environment for existence varies considerably with species. For example, dolphins and whales are completely dependent on the marine environment for all stages of their life; seals feed in the ocean but breed on land; and polar bears must feed on land. Twenty three percent of marine mammal species are threatened.

The cetaceans became aquatic around 50 million years ago (mya). Based on molecular and morphological research, the cetaceans genetically and morphologically fall firmly within the Artiodactyla (even-toed ungulates). The term "Cetartiodactyla" reflects the idea that whales evolved within the ungulates. The term was coined by merging the name for the two orders, Cetacea and Artiodactyla, into a single word. Under this definition, the closest living land relative of the whales and dolphins is thought to be the hippopotamuses.

Sirenians, the sea cows, became aquatic around 40 million years ago. The first appearance of sirenians in the fossil record was during the early Eocene, and by the late Eocene, sirenians had significantly diversified. Inhabitants of rivers, estuaries, and nearshore marine waters, they were able to spread rapidly. The most primitive sirenian, †Prorastomus, was found in Jamaica, unlike other marine mammals which originated from the Old World (such as cetaceans). The first known quadrupedal sirenian was †Pezosiren from the early Eocene. The earliest known sea cows, of the families †Prorastomidae and †Protosirenidae, were both confined to the Eocene, and were pig-sized, four-legged, amphibious creatures. The first members of Dugongidae appeared by the end of the Eocene. At this point, sea cows were fully aquatic.

Pinnipeds split from other caniforms 50 mya during the Eocene. Their evolutionary link to terrestrial mammals was unknown until the 2007 discovery of †Puijila darwini in early Miocene deposits in Nunavut, Canada. Like a modern otter, †Puijila had a long tail, short limbs and webbed feet instead of flippers. The lineages of Otariidae (eared seals) and Odobenidae (walrus) split almost 28 mya. Phocids (earless seals) are known to have existed for at least 15 mya, and molecular evidence supports a divergence of the Monachinae (monk seals) and Phocinae lineages 22 mya.

Fossil evidence indicates the sea otter (Enhydra) lineage became isolated in the North Pacific approximately two mya, giving rise to the now-extinct †Enhydra macrodonta and the modern sea otter, Enhydra lutris. The sea otter evolved initially in northern Hokkaidō and Russia, and then spread east to the Aleutian Islands, mainland Alaska, and down the North American coast. In comparison to cetaceans, sirenians, and pinnipeds, which entered the water approximately 50, 40, and 20 mya, respectively, the sea otter is a relative newcomer to marine life. In some respects though, the sea otter is more fully adapted to water than pinnipeds, which must haul out on land or ice to give birth.

Polar bears are thought to have diverged from a population of brown bears, Ursus arctos, that became isolated during a period of glaciation in the Pleistocene or from the eastern part of Siberia, (from Kamchatka and the Kolym Peninsula). The oldest known polar bear fossil is a 130,000 to

110,000-year-old jaw bone, found on Prince Charles Foreland in 2004. The mitochondrial DNA (mtDNA) of the polar bear diverged from the brown bear roughly 150,000 years ago. Further, some clades of brown bear, as assessed by their mtDNA, are more closely related to polar bears than to other brown bears, meaning that the polar bear might not be considered a species under some species concepts.

In general, terrestrial amniote invasions of the sea have become more frequent in the Cenozoic than they were in the Mesozoic. Factors contributing to this trend include the increasing productivity of near-shore marine environments, and the role of endothermy in facilitating this transition.

Marine mammals of varying sizes and shapes:

A polar bear (Ursus maritimus), a member of family Ursidae.

A sea otter (Enhydra lutris), a member of family Mustelidae.

California sea lions (Zalophus californianus), members of the family Otariidae.

A West Indian manatee (Trichechus manatus), a member of order Sirenia.

A common minke whale (Balaenoptera acutorostrata), a member of the order Cetartiodactyla.

Classification of Extant Species

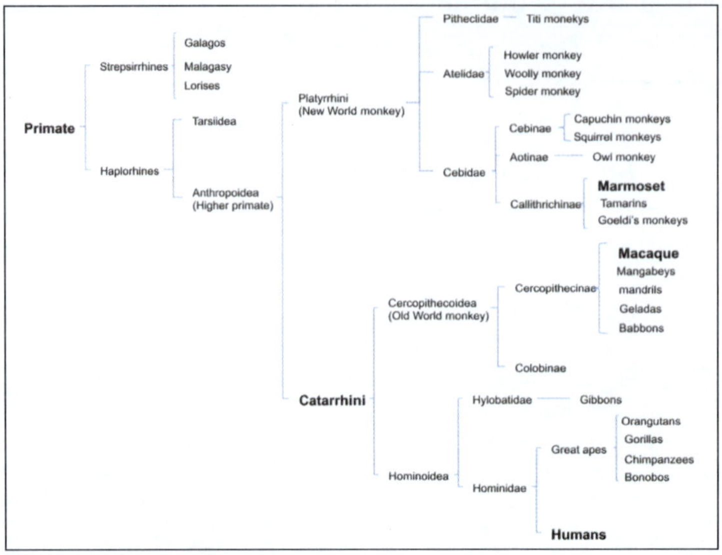

Order Cetartiodactyla:

- Suborder Whippomorpha.
- Family Balaenidae (right and bowhead whales), two genera and four species.
- Family Cetotheriidae (pygmy right whale), one species.
- Family Balaenopteridae (rorquals), two genera and eight species.
- Family Eschrichtiidae (gray whale), one species.
- Family Physeteridae (sperm whale), one species.
- Family Kogiidae (pygmy and dwarf sperm whales), one genus and two species.
- Family Monodontidae (narwhal and beluga), two genera and two species.
- Family Ziphiidae (beaked whales), six genera and 21 species.
- Family Delphinidae (oceanic dolphins), 17 genera and 38 species.
- Family Phocoenidae (porpoises), two genera and seven species.

Order Sirenia (sea cows):

- Family Trichechidae (manatees), two species.
- Family Dugongidae (dugongs), one species.

Order Carnivora (carnivores):

- Suborder Caniformia.
- Family Mustelidae, three species.
- Family Ursidae (bears), one species.

Suborder Pinnipedia (sealions, walruses, seals):

- Family Otariidae (eared seals), seven genera and 15 species.
- Family Odobenidae (walrus), one species.
- Family Phocidae (earless seals), 14 genera and 18 species.

Distribution and Habitat

Marine mammals are widely distributed throughout the globe, but their distribution is patchy and coincides with the productivity of the oceans. Species richness peaks at around 40° latitude, both north and south. This corresponds to the highest levels of primary production around North and South America, Africa, Asia and Australia. Total species range is highly variable for marine mammal species. On average most marine mammals have ranges which are equivalent or smaller than one-fifth of the Indian Ocean. The variation observed in range size is a result of the different ecological requirements of each species and their ability to cope with a broad range of environmental conditions. The high degree of overlap between marine mammal species richness and areas of human impact on the environment is of concern.

Most marine mammals, such as seals and sea otters, inhabit the coast. Seals, however, also use a number of terrestrial habitats, both continental and island. In temperate and tropical areas, they haul-out on to sandy and pebble beaches, rocky shores, shoals, mud flats, tide pools and in sea caves. Some species also rest on man-made structures, like piers, jetties, buoys and oil platforms. Seals may move further inland and rest in sand dunes or vegetation, and may even climb cliffs.:96 Most cetaceans live in the open ocean, and species like the sperm whale may dive to depths of −1,000 to −2,500 feet (−300 to −760 m) in search of food. Sirenians live in shallow coastal waters, usually living 30 feet (9.1 m) below sea level. However, they have been known to dive to −120 feet (−37 m) to forage deep-water seagrasses. Sea otters live in protected areas, such as rocky shores, kelp forests, and barrier reefs, although they may reside among drift ice or in sandy, muddy, or silty areas.

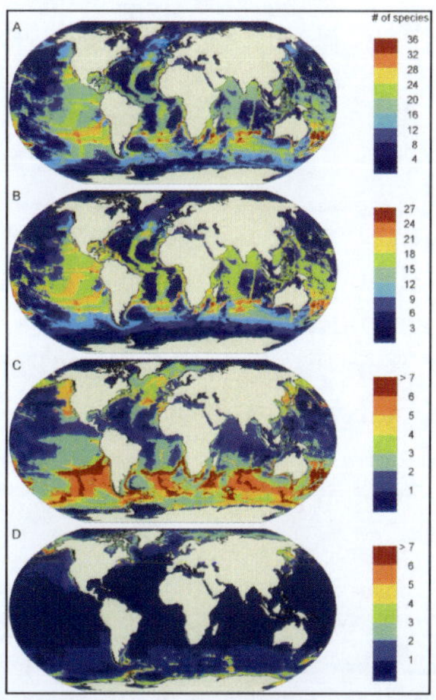

Marine mammal species richness: A) All species (n=115), B) Odontocetes (n=69), C) Mysticetes (n=14), D) Pinnipeds (n=32), based on data from 1990–1999.

Many marine mammals seasonally migrate. Annual ice contains areas of water that appear and disappear throughout the year as the weather changes, and seals migrate in response to these changes. In turn, polar bears must follow their prey. In Hudson Bay, James Bay, and some other areas, the ice melts completely each summer (an event often referred to as "ice-floe breakup"), forcing polar bears to go onto land and wait through the months until the next freeze-up. In the Chukchi and Beaufort seas, polar bears retreat each summer to the ice further north that remains frozen year-round. Seals may also migrate to other environmental changes, such as El Niño, and traveling seals may use various features of their environment to reach their destination including geomagnetic fields, water and wind currents, the position of the sun and moon and the taste and temperature of the water. Baleen whales famously migrate very long distances into tropical waters to give birth and raise young, possibly to prevent predation by killer whales. The gray whale has the longest recorded migration of any mammal, with one traveling 14,000 miles (23,000 km) from the Sea of Okhotsk to the Baja Peninsula. During the winter, manatees living at the northern end of their range migrate to warmer waters.

Adaptations

Marine mammals have a number of physiological and anatomical features to overcome the unique challenges associated with aquatic living. Some of these features are very species-specific. Marine mammals have developed a number of features for efficient locomotion such as torpedo-shaped bodies to reduce drag; modified limbs for propulsion and steering; tail flukes and dorsal fins for propulsion and balance. Marine mammals are adept at thermoregulation using dense fur or blubber, circulatory adjustments (counter-current heat exchangers); torpedo-shaped bodies, reduced appendages, and large size to prevent heat loss.

Marine mammals are able to dive for long periods of time. Both pinnipeds and cetaceans have large and complex blood vessel systems which serve to store oxygen to support deep diving. Other important reservoirs include muscles, blood, and the spleen which all have the capacity to hold a high concentration of oxygen. They are also capable of bradycardia (reduced heart rate), and vasoconstriction (shunting most of the oxygen to vital organs such as the brain and heart) to allow extended diving times and cope with oxygen deprivation. If oxygen is depleted, marine mammals can access substantial reservoirs of glycogen that support anaerobic glycolysis of the cells involved during conditions of systemic hypoxia associated with prolonged submersion.

Sound travels differently through water, and therefore marine mammals have developed adaptations to ensure effective communication, prey capture, and predator detection. The most notable adaptation is the development of echolocation in whales and dolphins. Toothed whales emit a focused beam of high-frequency clicks in the direction that their head is pointing. Sounds are generated by passing air from the bony nares through the phonic lips. These sounds are reflected by the dense concave bone of the cranium and an air sac at its base. The focused beam is modulated by a large fatty organ known as the 'melon'. This acts like an acoustic lens because it is composed of lipids of differing densities.

Marine mammals have evolved a wide variety of features for feeding, which are mainly seen in their dentition. For example, the cheek teeth of pinnipeds and odontocetes are specifically adapted to capture fish and squid. In contrast, baleen whales have evolved baleen plates to filter feed plankton and small fish from the water.

Polar bears, otters, and fur seals have long, oily, and waterproof fur in order to trap air to provide insulation. In contrast, other marine mammals–such as whales, dolphins, porpoises, manatees, dugongs, and walruses–have lost long fur in favor of a thick, dense epidermis and a thickened fat layer (blubber) to prevent drag. Wading and bottom-feeding animals (such as manatees) need to be heavier than water in order to keep contact with the floor or to stay submerged. Surface-living animals (such as sea otters) need the opposite, and free-swimming animals living in open waters (such as dolphins) need to be neutrally buoyant in order to be able to swim up and down the water column. Typically, thick and dense bone is found in bottom feeders and low bone density is associated with mammals living in deep water. Some marine mammals, such as polar bears and otters, have retained four weight-bearing limbs and can walk on land like fully terrestrial animals.

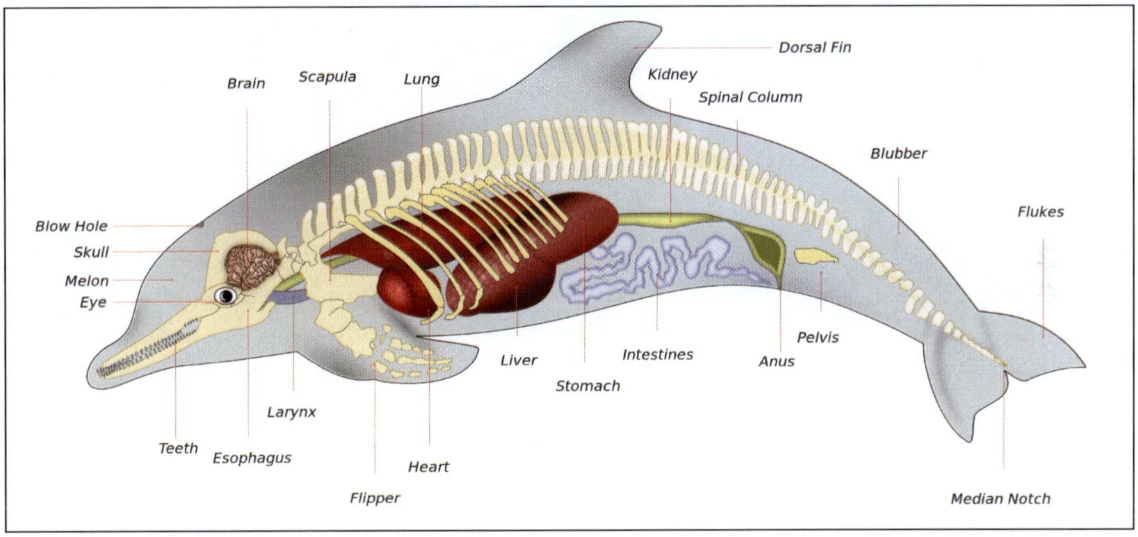

The anatomy of a dolphin showing its skeleton, major organs, and body shape.

Diet

All cetaceans are carnivorous and predatory. Toothed whales mostly feed on fish and cephalopods, followed by crustaceans and bivalves. Some may forage with other kinds of animals, such as other species of whales or certain species of pinnipeds. One common feeding method is herding, where a pod squeezes a school of fish into a small volume, known as a bait ball. Individual members then take turns plowing through the ball, feeding on the stunned fish. Coralling is a method where dolphins chase fish into shallow water to catch them more easily. Killer whales and bottlenose dolphins have also been known to drive their prey onto a beach to feed on it. Killer whales have been known to have paralyzed great white sharks and other sharks and rays by flipping it upside down. Other whales with a blunt snout and reduced dentition rely on suction feeding. Though carnivorous, they house gut flora similar to that of terrestrial herbivores, probably a remnant of their herbivorous ancestry.

Baleen whales use their baleen plates to sieve plankton, among others, out of the water; there are two types of methods: lunge-feeding and gulp-feeding. Lunge-feeders expand the volume of their jaw to a volume bigger than the original volume of the whale itself by inflating their mouth. This causes grooves on their throat to expand, increasing the amount of water the mouth can store. They ram a baitball at high speeds in order to feed, but this is only energy-effective when used against a large baitball. Gulp-feeders swim with an open mouth, filling it with water and prey. Prey must occur in sufficient numbers to trigger the whale's interest, be within a certain size range so that the baleen plates can filter it, and be slow enough so that it cannot escape.

Otters are the only marine animals that are capable of lifting and turning over rocks, which they often do with their front paws when searching for prey. The sea otter may pluck snails and other organisms from kelp and dig deep into underwater mud for clams. It is the only marine mammal that catches fish with its forepaws rather than with its teeth. Under each foreleg, sea otters have a loose pouch of skin that extends across the chest which they use to store collected food to bring to the surface. This pouch also holds a rock that is used to break open shellfish and clams, an example of tool use. The sea otters eat while floating on their backs, using their forepaws to tear food apart and bring to their mouths. Marine otters mainly feed on crustaceans and fish.

Pinnipeds mostly feed on fish and cephalopods, followed by crustaceans and bivalves, and then zooplankton and warm-blooded prey (like sea birds). Most species are generalist feeders, but a few are specialists. They typically hunt non-schooling fish, slow-moving or immobile invertebrates or endothermic prey when in groups. Solitary foraging species usually exploit coastal waters, bays and rivers. When large schools of fish or squid are available, pinnipeds hunt cooperatively in large groups, locating and herding their prey. Some species, such as California and South American sea lions, may forage with cetaceans and sea birds.

The polar bear is the most carnivorous species of bear, and its diet primarily consists of ringed (Pusa hispida) and bearded (Erignathus barbatus) seals. Polar bears hunt primarily at the interface between ice, water, and air; they only rarely catch seals on land or in open water. The polar bear's most common hunting method is still-hunting: The bear locates a seal breathing hole using its sense of smell, and crouches nearby for a seal to appear. When the seal exhales, the bear smells its breath, reaches into the hole with a forepaw, and drags it out onto the ice. The polar bear also hunts by stalking seals resting on the ice. Upon spotting a seal, it walks to within 100 yards (90 m), and then crouches. If the seal does not notice, the bear creeps to within 30 to 40 feet (9 to 10 m) of the seal and then suddenly rushes to attack. A third hunting method is to raid the birth lairs that female seals create in the snow. They may also feed on fish.

Sirenians are referred to as "sea cows" because their diet consists mainly of sea-grass. When eating, they ingest the whole plant, including the roots, although when this is impossible they feed on just the leaves. A wide variety of seagrass has been found in dugong stomach contents, and evidence exists they will eat algae when seagrass is scarce. West Indian manatees eat up to 60 different species of plants, as well as fish and small invertebrates to a lesser extent.

Killer whale hunting a Weddel seal.

Sea otters have dexterous hands which they use to smash sea urchins off rocks.

A dugong feeding on the sea-floor.

Keystone Species

Sea otters are a classic example of a keystone species; their presence affects the ecosystem more profoundly than their size and numbers would suggest. They keep the population of certain benthic (sea floor) herbivores, particularly sea urchins, in check. Sea urchins graze on the lower stems of kelp, causing the kelp to drift away and die. Loss of the habitat and nutrients provided by kelp forests leads to profound cascade effects on the marine ecosystem. North Pacific areas that do not have sea otters often turn into urchin barrens, with abundant sea urchins and no kelp forest. Reintroduction of sea otters to British Columbia has led to a dramatic improvement in the health of coastal ecosystems, and similar changes have been observed as sea otter populations recovered in the Aleutian and Commander Islands and the Big Sur coast of California However, some kelp forest ecosystems in California have also thrived without sea otters, with sea urchin populations apparently controlled by other factors. The role of sea otters in maintaining kelp forests has been observed to be more important in areas of open coast than in more protected bays and estuaries.

An apex predator affects prey population dynamics and defense tactics (such as camouflage). The polar bear is the apex predator within its range. Several animal species, particularly Arctic foxes (Vulpes lagopus) and glaucous gulls (Larus hyperboreus), routinely scavenge polar bear kills. The relationship between ringed seals and polar bears is so close that the abundance of ringed seals in some areas appears to regulate the density of polar bears, while polar bear predation in turn regulates density and reproductive success of ringed seals. The evolutionary pressure of polar bear predation on seals probably accounts for some significant differences between Arctic and Antarctic seals. Compared to the Antarctic, where there is no major surface predator, Arctic seals use more breathing holes per individual, appear more restless when hauled out on the ice, and rarely defecate on the ice. The fur of Arctic pups is white, presumably to provide camouflage from predators, whereas Antarctic pups all have dark fur.

Killer whales are apex predators throughout their global distribution, and can have a profound effect on the behavior and population of prey species. Their diet is very broad and they can feed on many vertebrates in the ocean including salmon, rays, sharks (even white sharks), large baleen whales, and nearly 20 species of pinniped. The predation of whale calves may be responsible for annual whale migrations to calving grounds in more tropical waters, where the population of killer whales is much lower than in polar waters. Prior to whaling, it is thought that great whales were a major food source; however, after their sharp decline, killer whales have since expanded their diet, leading to the decline of smaller marine mammals. A decline in Aleutian Islands sea otter populations in the 1990s was controversially attributed by some scientists to killer whale predation, although with no direct evidence. The decline of sea otters followed a decline in harbor seal and Steller sea lion populations, the killer whale's preferred prey, which in turn may be substitutes for their original prey, now reduced by industrial whaling.

Whale Pump

A 2010 study considered whales to be a positive influence to the productivity of ocean fisheries, in what has been termed a "whale pump". Whales carry nutrients such as nitrogen from the depths back to the surface. This functions as an upward biological pump, reversing an earlier presumption that whales accelerate the loss of nutrients to the bottom. This nitrogen input in the Gulf of Maine is more than the input of all rivers combined emptying into the gulf, some 25,000 short tons (23,000 t) each year. Whales defecate at the ocean's surface; their excrement is important for fisheries because it is rich in iron and nitrogen. The whale feces are liquid and instead of sinking, they stay at the surface where phytoplankton feed off it.

Upon death, whale carcasses fall to the deep ocean and provide a substantial habitat for marine life. Evidence of whale falls in present-day and fossil records shows that deep sea whale falls support a rich assemblage of creatures, with a global diversity of 407 species, comparable to other neritic biodiversity hotspots, such as cold seeps and hydrothermal vents. Deterioration of whale carcasses happens though a series of three stages. Initially, moving organisms, such as sharks and hagfish, scavenge soft tissue at a rapid rate over a period of months to as long as two years. This is followed by the colonization of bones and surrounding sediments (which contain organic matter) by enrichment opportunists, such as crustaceans and polychaetes, throughout a period of years. Finally, sulfophilic bacteria reduce the bones releasing hydrogen sulphide enabling the growth of chemoautotrophic organisms, which in turn, support other organisms such as mussels, clams, limpets, and sea snails. This stage may last for decades and supports a rich assemblage of species, averaging 185 species per site.

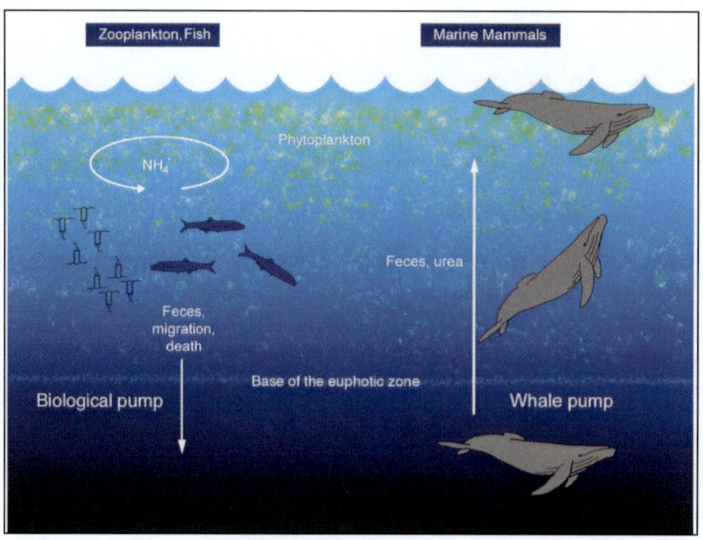

"Whale pump" – the role played by whales in recycling ocean nutrients.

Threats

Exploitation

Marine mammals were hunted by coastal aboriginal humans historically for food and other resources. These subsistence hunts still occur in Canada, Greenland, Indonesia, Russia, the United States, and several nations in the Caribbean. The effects of these are only localized, as hunting efforts were on a relatively small scale. Commercial hunting took this to a much greater scale and marine mammals were heavily exploited. This led to the extinction of the †Steller's sea cow (Hydrodamalis gigas), †sea mink (Neovison macrodon), †Japanese sea lion (Zalophus japonicus), and the †Caribbean monk seal (Neomonachus tropicalis). Today, populations of species that were historically hunted, such as blue whales (Balaenoptera musculus) and the North Pacific right whale (Eubalaena japonica), are much lower than their pre-whaling levels. Because whales generally have slow growth rates, are slow to reach sexual maturity, and have a low reproductive output, population recovery has been very slow.

A number of whales are still subject to direct hunting, despite the 1986 moratorium ban on whaling set under the terms of the International Whaling Commission (IWC). There are only two nations remaining which sanction commercial whaling: Norway, where several hundred common minke whales are harvested each year; and Iceland, where quotas of 150 fin whales and 100 minke whales per year are set. Japan also harvests several hundred Antarctic and North Pacific minke whales each year, ostensibly for scientific research in accordance with the moratorium. However, the illegal trade of whale and dolphin meat is a significant market in Japan and some countries.

The most profitable furs in the fur trade were those of sea otters, especially the northern sea otter which inhabited the coastal waters between the Columbia River to the south and Cook Inlet to the north. The fur of the Californian southern sea otter was less highly prized and thus less profitable. After the northern sea otter was hunted to local extinction, maritime fur traders shifted to California until the southern sea otter was likewise nearly extinct. The British and American maritime fur traders took their furs to the Chinese port of Guangzhou (Canton), where they worked within the established Canton System. Furs from Russian America were mostly sold to China via the Mongolian trading town of Kyakhta, which had been opened to Russian trade by the 1727 Treaty of Kyakhta.

Men killing northern fur seals on Saint Paul Island, Alaska in the 1890s.

Commercial sealing was historically just as important as the whaling industry. Exploited species included harp seals, hooded seals, Caspian seals, elephant seals, walruses and all species of fur seal. The scale of seal harvesting decreased substantially after the 1960s, after the Canadian government reduced the length of the hunting season and implemented measures to protect adult females. Several species that were commercially exploited have rebounded in numbers; for example, Antarctic fur seals may be as numerous as they were prior to harvesting. The northern elephant seal was hunted to near extinction in the late 19th century, with only a small population remaining on Guadalupe Island. It has since recolonized much of its historic range, but has a population bottleneck. Conversely, the Mediterranean monk seal was extirpated from much of its former range, which stretched from the Mediterranean to the Black Sea and northwest Africa, and only remains in the northeastern Mediterranean and some parts of northwest Africa.

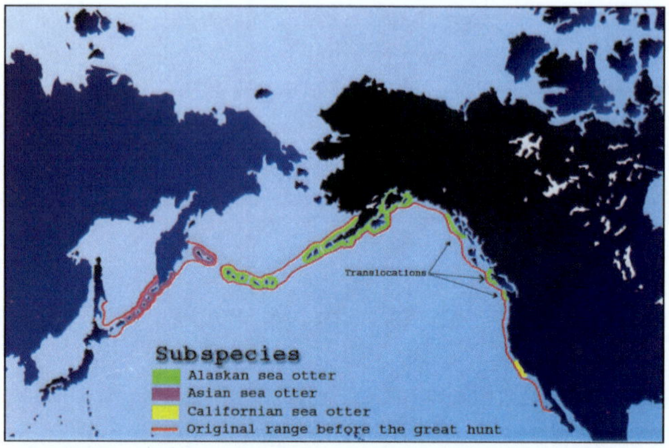

Historical and modern range of northern sea otters.

Polar bears can be hunted for sport in Canada with a special permit and accompaniment by a local guide. This can be an important source of income for small communities, as guided hunts bring in more income than selling the polar bear hide on markets. The United States, Russia, Norway, Greenland, and Canada allow subsistence hunting, and Canada distributes hunting permits to indigenous communities. The selling of these permits is a main source of income for many of these communities. Their hides can be used for subsistence purposes, kept as hunting trophies, or can be bought in markets.

Ocean Traffic and Fisheries

By-catch is the incidental capture of non-target species in fisheries. Fixed and drift gill nets cause the highest mortality levels for both cetaceans and pinnipeds, however, entanglements in long lines, mid-water trawls, and both trap and pot lines are also common. Tuna seines are particularly problematic for entanglement by dolphins. By-catch affects all cetaceans, both small and big, in all habitat types. However, smaller cetaceans and pinnipeds are most vulnerable as their size means that escape once they are entangled is highly unlikely and they frequently drown. While larger cetaceans are capable of dragging nets with them, the nets sometimes remain tightly attached to the individual and can impede the animal from feeding sometimes leading to starvation. Abandoned or lost nets and lines cause mortality through ingestion or entanglement. Marine mammals also get entangled in aquaculture nets, however, these are rare events and not prevalent enough to impact populations.

Vessel strikes cause death for a number of marine mammals, especially whales. In particular, fast commercial vessels such as container ships can cause major injuries or death when they collide with marine mammals. Collisions occur both with large commercial vessels and recreational boats and cause injury to whales or smaller cetaceans. The critically endangered North Atlantic right whale is particularly affected by vessel strikes. Tourism boats designed for whale and dolphin watching can also negatively impact on marine mammals by interfering with their natural behavior.

The fishery industry not only threatens marine mammals through by-catch, but also through competition for food. Large scale fisheries have led to the depletion of fish stocks that are important prey species for marine mammals. Pinnipeds have been especially affected by the direct loss of food supplies and in some cases the harvesting of fish has led to food shortages or dietary deficiencies, starvation of young, and reduced recruitment into the population. As the fish stocks have been depleted, the competition between marine mammals and fisheries has sometimes led to conflict. Large-scale culling of populations of marine mammals by commercial fishers has been initiated in a number of areas in order to protect fish stocks for human consumption.

The remains of a North Atlantic right whale after it collided with a ship propeller.

Shellfish aquaculture takes up space so in effect creates competition for space. However, there is little direct competition for aquaculture shellfish harvest. On the other hand, marine mammals regularly take finfish from farms, which creates significant problems for marine farmers. While there are usually legal mechanisms designed to deter marine mammals, such as anti-predator nets or harassment devices, individuals are often illegally shot.

Habitat Loss and Degradation

Habitat degradation is caused by a number of human activities. Marine mammals that live in coastal environments are most likely to be affected by habitat degradation and loss. Developments such as sewage marine outfalls, moorings, dredging, blasting, dumping, port construction, hydroelectric projects, and aquaculture both degrade the environment and take up valuable habitat. For example, extensive shellfish aquaculture takes up valuable space used by coastal marine mammals for important activities such as breeding, foraging and resting.

Contaminants that are discharged into the marine environment accumulate in the bodies of marine mammals when they are stored unintentionally in their blubber along with energy. Contaminants that are found in the tissues of marine mammals include heavy metals, such as mercury and lead, but also organochlorides and polycyclic aromatic hydrocarbons. For example, these can cause disruptive effects on endocrine systems; impair the reproductive system, and lower the immune system of individuals, leading to a higher number of deaths. Other pollutants such as oil, plastic debris and sewage threaten the livelihood of marine mammals.

Noise pollution from anthropogenic activities is another major concern for marine mammals. This is a problem because underwater noise pollution interferes with the abilities of some marine mammals to communicate, and locate both predators and prey. Underwater explosions are used for a variety of purposes including military activities, construction and oceanographic or geophysical research. They can cause injuries such as hemorrhaging of the lungs, and contusion and ulceration of the gastrointestinal tract. Underwater noise is generated from shipping, the oil and gas industry, research, and military use of sonar and oceanographic acoustic experimentation. Acoustic harassment devices and acoustic deterrent devices used by aquaculture facilities to scare away marine mammals emit loud and noxious underwater sounds.

Two changes to the global atmosphere due to anthropogenic activity threaten marine mammals. The first is increases in ultraviolet radiation due to ozone depletion, and this mainly affects the Antarctic and other areas of the southern hemisphere. An increase in ultraviolet radiation has the capacity to decrease phytoplankton abundance, which forms the basis of the food chain in the ocean. The second effect of global climate change is global warming due to increased carbon dioxide levels in the atmosphere. Raised sea levels, sea temperature and changed currents are expected to affect marine mammals by altering the distribution of important prey species, and changing the suitability of breeding sites and migratory routes. The Arctic food chain would be disrupted by the near extinction or migration of polar bears. Arctic sea ice is the polar bear's habitat. It has been declining at a rate of 13% per decade because the temperature is rising at twice the rate of the rest of the world. By the year 2050, up to two-thirds of the world's polar bears may vanish if the sea ice continues to melt at its current rate.

A study by evolutionary biologists at the University of Pittsburgh showed that the ancestors of many marine mammals stopped producing a certain enzyme that today protects against some neurotoxic chemicals called organophosphates, including those found in the widely used pesticides chlorpyrifos and diazinon. Marine mammals may be increasingly exposed to these compounds due to agricultural runoff reaching the world's oceans.

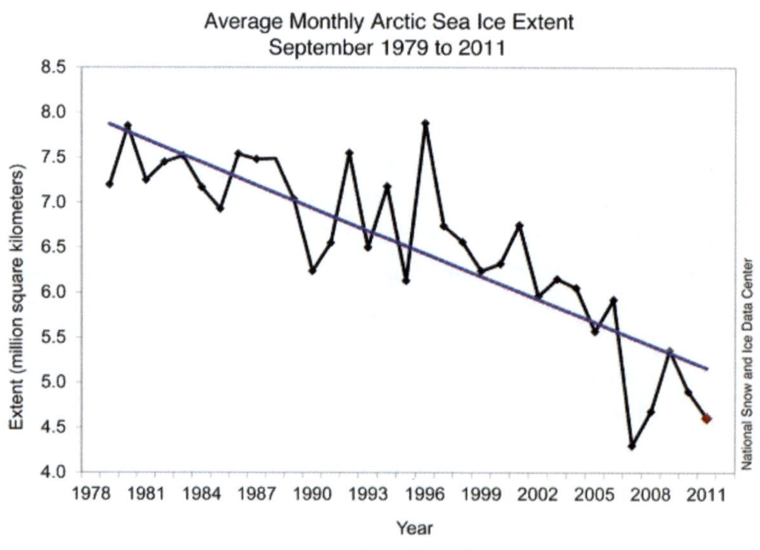

Map from the U.S. Geological Survey shows projected changes in polar bear habitat from 2005 to 2095. Red areas indicate loss of optimal polar bear habitat; blue areas indicate gain.

Protection

The Marine Mammal Protection Act of 1972 (MMPA) was passed on October 21, 1972 under president Richard Nixon to prevent the further depletion and possible extinction of marine mammal stocks. It prohibits the taking ("the act of hunting, killing, capture, and harassment of any marine mammal; or, the attempt at such") of any marine mammal without a permit issued by the Secretary. Authority to manage the MMPA was divided between the Secretary of the Interior through the U.S. Fish and Wildlife Service (Service), and the Secretary of Commerce, which is delegated to the National Oceanic and Atmospheric Administration (NOAA). The Marine Mammal Commission (MMC) was established to review existing policies and make recommendations to the Service and NOAA to better implement the MMPA. The Service is responsible for ensuring the protection of sea otters and marine otters, walruses, polar bears, the three species of manatees, and dugongs; and NOAA was given responsibility to conserve and manage pinnipeds (excluding walruses) and cetaceans.

The Act was updated on 1 January 2016 with a clause banning "the import of fish from fisheries that cannot prove they meet US standards for protecting marine mammals." The requirement to show that protection standards are met is hoped to compel countries exporting fish to the US to more strictly control their fisheries that no protected marine mammals are adversely affected by fishing.

The 1979 Convention on the Conservation of Migratory Species of Wild Animals (CMS) is the only global organization that conserves a broad range of animals, of which includes marine mammals. Of the agreements made, three of them deal with the conservation of marine mammals: ACCOBAMS, ASCOBANS, and the Wadden Sea Agreement. In 1982, the United Nations Convention on the Law of the Sea (LOSC) adopted a pollution prevention approach to conservation, which many other conventions at the time also adopted.

The Agreement on the Conservation of Cetaceans in the Black Sea, Mediterranean Sea and contiguous Atlantic area (ACCOBAMS), founded in 1996, specifically protects cetaceans in the Mediterranean area, and "maintains a favorable status", a direct action against whaling. There are 23 member states. The Agreement on the Conservation of Small Cetaceans of the Baltic and North Seas (ASCOBANS) was adopted alongside ACCOBAMS to establish a special protection area for Europe's increasingly threatened cetaceans. Other anti-whaling efforts include a ten-year moratorium in 1986 by the IWC on all whaling, and an environmental agreement (a type of international law) the International Convention for the Regulation of Whaling which controlled commercial, scientific, and subsistence whaling.

The Agreement on the Conservation of Seals in the Wadden Sea, enforced in 1991, prohibits the killing or harassment of seals in the Wadden Sea, specifically targeting the harbor seal population.

The 1973 Agreement on the Conservation of Polar Bears between Canada, Denmark (Greenland), Norway (Svalbard), the United States, and the Soviet Union outlawed the unregulated hunting of polar bears from aircraft and icebreakers, as well as protecting migration, feeding, and hibernation sites.

Various non-governmental organizations participate in marine conservation activism, wherein they draw attention to and aid in various problems in marine conservation, such as pollution, whaling, bycatch, and so forth. Notable organizations include the Greenpeace who focus on overfishing and whaling among other things, and Sea Shepherd Conservation Society who are known for taking direct-action tactics to expose illegal activity.

Signatory countries of the International Whaling Commission (IWC).

As Food

For thousands of years, indigenous peoples of the Arctic have depended on whale meat. The meat is harvested from legal, non-commercial hunts that occur twice a year in the spring and autumn. The meat is stored and eaten throughout the winter. The skin and blubber (muktuk) taken from the bowhead, beluga, or narwhal is also valued, and is eaten raw or cooked. Whaling has also been practiced in the Faroe Islands in the North Atlantic since about the time of the first Norse settlements on the islands. Around 1000 long-finned pilot whales are still killed annually, mainly during the summer. Today, dolphin meat is consumed in a small number of countries worldwide, which include Japan and Peru. In some parts of the world, such as Taiji, Japan and the Faroe Islands, dolphins are traditionally considered food, and are killed in harpoon or drive hunts.

There have been human health concerns associated with the consumption of dolphin meat in Japan after tests showed that dolphin meat contained high levels of methylmercury. There are no known cases of mercury poisoning as a result of consuming dolphin meat, though the government continues to monitor people in areas where dolphin meat consumption is high. The Japanese government recommends that children and pregnant women avoid eating dolphin meat on a regular basis. Similar concerns exist with the consumption of dolphin meat in the Faroe Islands, where prenatal exposure to methylmercury and PCBs primarily from the consumption of pilot whale meat has resulted in neuropsychological deficits amongst children.

The Faroe Islands population was exposed to methylmercury largely from contaminated pilot whale meat, which contained very high levels of about 2 mg methylmercury/kg. However, the Faroe Islands populations also eat significant amounts of fish. The study of about 900 Faroese children showed that prenatal exposure to methylmercury resulted in neuropsychological deficits at 7 years of age.

Ringed seals were once the main food staple for the Inuit. They are still an important food source for the people of Nunavut and are also hunted and eaten in Alaska. Seal meat is an important source of food for residents of small coastal communities. The seal blubber is used to make seal oil, which is marketed as a fish oil supplement. In 2001, two percent of Canada's raw seal oil was processed and sold in Canadian health stores.

Pilot whale meat (bottom), blubber (middle) and dried fish (left) with potatoes, Faroe Islands.

Aquariums

Cetaceans

Various species of dolphins are kept in captivity. These small cetaceans are more often than not kept in theme parks and dolphinariums, such as SeaWorld. Bottlenose dolphins are the most common species of dolphin kept in dolphinariums as they are relatively easy to train and have a long lifespan in captivity. Hundreds of bottlenose dolphins live in captivity across the world, though exact numbers are hard to determine. The dolphin "smile" makes them popular attractions, as this is a welcoming facial expression in humans; however the smile is due to a lack of facial muscles and subsequent lack of facial expressions.

Performing killer whale at SeaWorld San Diego, 2009.

Organizations such as World Animal Protection and the Whale and Dolphin Conservation campaign against the practice of keeping cetaceans, particularly killer whales, in captivity. In captivity, they often develop pathologies, such as the dorsal fin collapse seen in 60–90% of male killer whales. Captives have vastly reduced life expectancies, on average only living into their 20s. In the wild, females who survive infancy live 46 years on average, and up to 70–80 years in rare cases. Wild males who survive infancy live 31 years on average, and up to 50–60 years. Captivity usually bears little resemblance to wild habitat, and captive whales' social groups are foreign to those found in the wild. Captive life is also stressful due the requirement to perform circus tricks that are not part of wild killer whale behavior, as well as restricting pool size. Wild killer whales may travel up to 100 miles (160 km) in a day, and critics say the animals are too big and intelligent to be suitable for captivity. Captives occasionally act aggressively towards themselves, their tankmates, or humans, which critics say is a result of stress. Dolphins are often trained to do several anthropomorphic behaviors, including waving and kissing—behaviors wild dolphins would rarely do.

Pinnipeds

The large size and playfulness of pinnipeds make them popular attractions. Some exhibits have rocky backgrounds with artificial haul-out sites and a pool, while others have pens with small rocky, elevated shelters where the animals can dive into their pools. More elaborate exhibits contain deep pools that can be viewed underwater with rock-mimicking cement as haul-out areas. The most common pinniped species kept in captivity is the California sea lion as it is abundant and easy to train. These animals are used to perform tricks and entertain visitors. Other species popularly kept in captivity include the grey seal and harbor seal. Larger animals like walruses and Steller sea lions are much less common. Pinnipeds are popular attractions because they are "disneyfied", and, consequently, people often anthropomorphize them with a curious, funny, or playful nature.

Some organizations, such as the Humane Society of the United States and World Animal Protection, object to keeping pinnipeds and other marine mammals in captivity. They state that the exhibits could not be large enough to house animals that have evolved to be migratory, and a pool could never replace the size and biodiversity of the ocean. They also oppose using sea lions for entertainment, claiming the tricks performed are "exaggerated variations of their natural behaviors" and distract the audience from the animal's unnatural environment.

Otter

Sea otters can do well in captivity, and are featured in over 40 public aquariums and zoos. The Seattle Aquarium became the first institution to raise sea otters from conception to adulthood with the birth of Tichuk in 1979, followed by three more pups in the early 1980s. In 2007, a YouTube video of two cute sea otters holding paws drew 1.5 million viewers in two weeks, and had over 20 million views as of January 2015. Filmed five years previously at the Vancouver Aquarium, it was YouTube's most popular animal video at the time, although it has since been surpassed. Otters are often viewed as having a "happy family life", but this is an anthropomorphism.

A sea lion trained to balance a ball on its nose.

Sirenians

The oldest manatee in captivity was Snooty, at the South Florida Museum's Parker Manatee Aquarium in Bradenton, Florida. Born at the Miami Aquarium and Tackle Company on July 21, 1948, Snooty was one of the first recorded captive manatee births. He was raised entirely in captivity, and died at the age of 69. Manatees can also be viewed in a number of European zoos, such as the Tierpark in Berlin, the Nuremberg Zoo, in ZooParc de Beauval in France, and in the Aquarium of Genoa in Italy. The River Safari at Singapore features seven of them.

Military

Bottlenose dolphins and California sea lions are used in the United States Navy Marine Mammal Program (NMMP) to detect mines, protect ships from enemy soldiers, and recover objects. The Navy has never trained attack dolphins, as they would not be able to discern allied soldiers from enemy soldiers. There were five marine mammal teams, each purposed for one of the three tasks: MK4 (dolphins), MK5 (sea lions), MK6 (dolphins and sea lions), MK7 (dolphins), and MK8 (dolphins); MK is short for mark. The dolphin teams were trained to detect and mark mines either attached to the seafloor or floating in the water column, because dolphins can use their echolocative abilities to detect mines. The sea lion team retrieved test equipment such as fake mines or bombs dropped from planes usually out of reach of divers who would have to make multiple dives. MK6 protects harbors and ships from enemy divers, and was operational in the Gulf War and Vietnam War. The dolphins would swim up behind enemy divers and attach a buoy to their air tank, so that they would float to the surface and alert nearby Navy personnel. Sea lions would hand-cuff the enemy, and try to outmaneuver their counter-attacks.

A dolphin wearing a locating pinger, performing mine clearance work in the Iraq War.

The use of marine mammals by the Navy, even in accordance with the Navy's policy, continues to meet opposition. The Navy's policy says that only positive reinforcement is to be used while training the military dolphins, and that they be cared for in accordance with accepted standards in animal care. The inevitable stresses involved in training are topics of controversy, as their treatment is unlike the animals' natural lifestyle, especially towards their confined spaces when not training. There is also controversy over the use of muzzles and other inhibitors, which prevent the dolphins from foraging for food while working. The Navy states that this is to prevent them from ingesting harmful objects, but conservation activists say this is done to reinforce the trainers' control over the dolphins, who hand out food rewards. The means of transportation is also an issue for conservation activists, since they are hauled in dry carriers, and switching tanks and introducing the dolphin to new dolphins is potentially dangerous as they are territorial.

Marine Worm

A marine worm is a worm that lives in the ocean, and there are myriad species of them. They come in various sizes, forms, and colors. Some worms are stationary, while others move about. Some worms are parasitic, some are predatory, and still others are plant-eaters. Many worms also possess the ability of regeneration if part of their body is lost, and some are capable of living in the harshest of environments. They all have something in common, though: they are an integral part of the ocean's ecosystem.

Christmas Tree Worm

This small stationary worm looks much like an anemone, and makes its home on coral mounds. It is beautifully colored and has a conical, swirled shape that comes from crowns consisting of feathery tentacles protruding from a tube-like body. It feeds on passing particles of plant matter and plankton, which are transferred from the water column to the mouth through the appendages on its crowns. These appendages, called radioles, also facilitate respiration, much like gills on fish.

Bearded Fireworm

This flat, segmented worm is a type of bristleworm, which is identified by the hollow, venom-filled bristles that run along both of its sides. This predatory worm is fantastically colored, with long bristles that are lost when the worm is disturbed. When handled, these bristles inflict a horrible burning sensation, giving the fireworm its very literal name.

Peanut Worm

This organic particle-eating worm is able to retract its proboscis into its body, much like the finger of a glove. The resulting shape when retracted is how the peanut worm got its name. The mouth part is covered in tentacles, which it uses to collect its food.

Bootlace Worm

This unattractive, long worm is part of the ribbon worm species and can reach lengths of approximately 197 feet. However, it is quite slender in diameter. This worm is deep brown in color and feeds on other worms, which it pulls inside of its body with its proboscis. Touching this worm is not advised, since its mucus contains a neurotoxin that serves as its defense against predators.

Spaghetti Worm

This worm is named for the tentacles at its mouth, which can be six times its body length. It uses these tentacles to grab passing food particles. If one of these tentacles is bitten or pulled off, it will grow back. Although there is some color variation, this worm is usually subtly colored.

Marine Invertebrates

Marine invertebrates are the invertebrates that live in marine habitats. Invertebrate is a blanket term that includes all animals apart from the vertebrate members of the chordate phylum. Invertebrates lack a vertebral column, and some have evolved a shell or a hard exoskeleton. As on land and in the air, marine invertebrates have a large variety of body plans, and have been categorised into over 30 phyla. They make up most of the macroscopic life in the oceans.

Ernst Haeckel's 96th plate, showing some marine invertebrates. Marine invertebrates have a large variety of body plans, which are currently categorised into over 30 phyla.

Classification

Invertebrates are grouped into different phyla. Informally phyla can be thought of as a way of grouping organisms according to their body plan. A body plan refers to a blueprint which describes the shape or morphology of an organism, such as its symmetry, segmentation and the disposition of its appendages. The idea of body plans originated with vertebrates, which were grouped into one phylum. But the vertebrate body plan is only one of many, and invertebrates consist of many phyla or body plans. The history of the discovery of body plans can be seen as a movement from a worldview centred on vertebrates, to seeing the vertebrates as one body plan among many. Among the pioneering zoologists, Linnaeus identified two body plans outside the vertebrates; Cuvier identified three; and Haeckel had four, as well as the Protista with eight more, for a total of twelve. For comparison, the number of phyla recognised by modern zoologists has risen to 35.

Historically body plans were thought of as having evolved in rapidly during the Cambrian explosion, but a more nuanced understanding of animal evolution suggests a gradual development of body plans throughout the early Palaeozoic and beyond. More generally a phylum can be defined in two ways: as described above, as a group of organisms with a certain degree of morphological or developmental similarity (the phenetic definition), or a group of organisms with a certain degree of evolutionary relatedness (the phylogenetic definition).

As on land and in the air, invertebrates make up a great majority of all macroscopic life, as the vertebrates makes up a subphylum of one of over 30 known animal phyla, making the term almost meaningless for taxonomic purpose. Invertebrate sea life includes the following groups, some of which are phyla:

- Acoela, among the most primitive bilateral animals.
- Annelida, (polychaetes and sea leeches).
- Brachiopoda, marine animals that have hard "valves" (shells) on the upper and lower surfaces .
- Bryozoa, also known as moss animals or sea mats.
- Chaetognatha, commonly known as arrow worms, are a phylum of predatory marine worms that are a major component of plankton.
- Cephalochordata represented in the modern oceans by the lancelets (also known as Amphioxus).
- Cnidaria, such as jellyfish, sea anemones, and corals.
- Crustacea, including lobsters, crabs, shrimp, crayfish, barnacles, hermit crabs, mantis shrimps, and copepods.
- Ctenophora, also known as comb jellies, the largest animals that swim by means of cilia.
- Echinodermata, including sea stars, brittle stars, sea urchins, sand dollars, sea cucumbers, crinoids, and sea daisies.
- Echiura, also known as spoon worms.
- Gnathostomulids, slender to thread-like worms, with a transparent body that inhabit sand and mud beneath shallow coastal waters.
- Gastrotricha, often called hairy backs, found mostly interstitially in between sediment particles.
- Hemichordata, includes acorn worms, solitary worm-shaped organisms.
- Kamptozoa, goblet-shaped sessile aquatic animals, with relatively long stalks and a "crown" of solid tentacles, also called Entoprocta.
- Kinorhyncha, segmented, limbless animals, widespread in mud or sand at all depths, also called mud dragons.
- Loricifera, very small to microscopic marine sediment-dwelling animals only discovered in 1983.
- Mollusca, including shellfish, squid, octopus, whelks, Nautilus, cuttlefish, nudibranchs, scallops, sea snails, Aplacophora, Caudofoveata, Monoplacophora, Polyplacophora, and Scaphopoda.

- Myzostomida, a taxonomic group of small marine worms which are parasitic on crinoids or "sea lilies".
- Nemertinea, also known as "ribbon worms" or "proboscis worms".
- Orthonectida, a small phylum of poorly known parasites of marine invertebrates that are among the simplest of multi-cellular organisms.
- Phoronida, a phylum of marine animals that filter-feed with a lophophore (a "crown" of tentacles), and build upright tubes of chitin to support and protect their soft bodies.
- Placozoa, small, flattened, multicellular animals around 1 millimetre across and the simplest in structure. They have no regular outline, although the lower surface is somewhat concave, and the upper surface is always flattened.
- Porifera (sponges), multicellular organisms that have bodies full of pores and channels allowing water to circulate through them.
- Priapulida, or penis worms, are a phylum of marine worms that live marine mud. They are named for their extensible spiny proboscis, which, in some species, may have a shape like that of a human penis.
- Pycnogonida, also called sea spiders, are unrelated to spiders, or even to arachnids which they resemble.
- Sipunculida, also called peanut worms, is a group containing 144–320 species (estimates vary) of bilaterally symmetrical, unsegmented marine worms.
- Tunicata, also known as sea squirts or sea pork, are filter feeders attached to rocks or similarly suitable surfaces on the ocean floor.
- Some flatworms of the classes Turbellaria and Monogenea.
- Xenoturbella, a genus of bilaterian animals that contains only two marine worm-like species;
- Xiphosura, includes a large number of extinct lineages and only four recent species in the family Limulidae, which include the horseshoe crabs.
- Arthropods total about 1,113,000 described extant species, molluscs about 85,000 and chordates about 52,000.

Marine Sponges

Sponges are animals of the phylum Porifera. They are multicellular organisms that have bodies full of pores and channels allowing water to circulate through them, consisting of jelly-like mesohyl sandwiched between two thin layers of cells. They have unspecialized cells that can transform into other types and that often migrate between the main cell layers and the mesohyl in the process. Sponges do not have nervous, digestive or circulatory systems. Instead, most rely on maintaining a constant water flow through their bodies to obtain food and oxygen and to remove wastes.

Sponges are similar to other animals in that they are multicellular, heterotrophic, lack cell walls and produce sperm cells. Unlike other animals, they lack true tissues and organs, and have no body symmetry. The shapes of their bodies are adapted for maximal efficiency of water flow through the central cavity, where it deposits nutrients, and leaves through a hole called the osculum. Many sponges have internal skeletons of spongin and/or spicules of calcium carbonate or silicon dioxide. All sponges are sessile aquatic animals. Although there are freshwater species, the great majority are marine (salt water) species, ranging from tidal zones to depths exceeding 8,800 m (5.5 mi).

Barrel sponge.

While most of the approximately 5,000–10,000 known species feed on bacteria and other food particles in the water, some host photosynthesizing micro-organisms as endosymbionts and these alliances often produce more food and oxygen than they consume. A few species of sponge that live in food-poor environments have become carnivores that prey mainly on small crustaceans.

Branching vase sponge.

Stove-pipe sponge.

Linnaeus mistakenly identified sponges as plants in the order Algae. For a long time thereafter sponges were assigned to a separate subkingdom, Parazoa (meaning beside the animals). They are now classified as a paraphyletic phylum from which the higher animals have evolved.

Marine Cnidarians

Cnidarians are distinguished by the presence of stinging cells, specialized cells that they use mainly for capturing prey. Cnidarians include corals, sea anemones, jellyfish and hydrozoans. They form a phylum containing over 10,000 species of animals found exclusively in aquatic (mainly marine) environments. Their bodies consist of mesoglea, a non-living jelly-like substance, sandwiched between two layers of epithelium that are mostly one cell thick. They have two basic body forms: swimming medusae and sessile polyps, both of which are radially symmetrical with mouths surrounded by tentacles that bear cnidocytes. Both forms have a single orifice and body cavity that are used for digestion and respiration.

Yet the starlet sea anemone contains the same genes as those that form the vertebrate head.

Sea anemones are common in tidepools.

Fossil cnidarians have been found in rocks formed about 580 million years ago. Fossils of cnidarians that do not build mineralized structures are rare. Scientists currently think cnidarians, ctenophores and bilaterians are more closely related to calcareous sponges than these are to other sponges, and that anthozoans are the evolutionary "aunts" or "sisters" of other cnidarians, and the most closely related to bilaterians.

Close up of polyps on the surface of a coral, waving their tentacles.

If an island sinks below the sea, coral growth can keep pace with the rising water and form an atoll.

Cnidarians are the simplest animals in which the cells are organised into tissues. The starlet sea anemone is used as a model organism in research. It is easy to care for in the laboratory and a protocol has been developed which can yield large numbers of embryos on a daily basis. There is a remarkable degree of similarity in the gene sequence conservation and complexity between the sea anemone and vertebrates. In particular, genes concerned in the formation of the head in vertebrates are also present in the anemone.

Their tentacles sting and paralyse small fish.

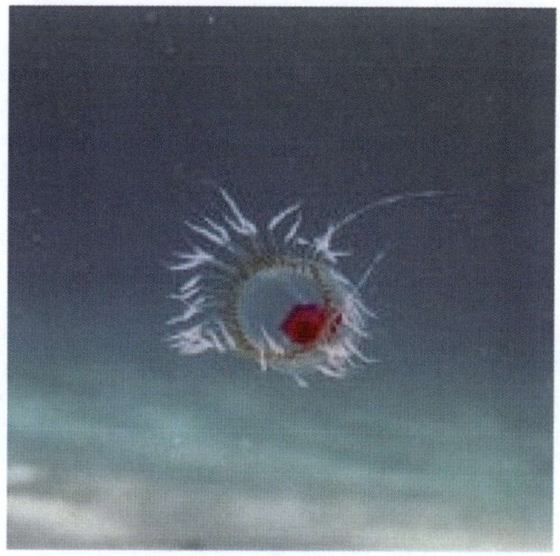

Turritopsis dohrnii achieves biological immortality by transferring its cells back to childhood.

Marine Worms

Worms (Old English for serpent) typically have long cylindrical tube-like bodies and no limbs. Marine worms vary in size from microscopic to over 1 metre (3.3 ft) in length for some marine polychaete worms (bristle worms) and up to 58 metres (190 ft) for the marine nemertean worm (bootlace worm). Some marine worms occupy a small variety of parasitic niches, living inside the bodies of other animals, while others live more freely in the marine environment or by burrowing underground.

Different groups of marine worms are related only distantly, so they are found in several different phyla such as the Annelida (segmented worms), Chaetognatha (arrow worms), Hemichordata, and Phoronida (horseshoe worms). Many of these worms have specialized tentacles used for exchanging oxygen and carbon dioxide and also may be used for reproduction. Some marine worms are tube worms, such as the giant tube worm which lives in waters near underwater volcanoes and can withstand temperatures up to 90 degrees Celsius.

Platyhelminthes (flatworms) form another worm phylum which includes a class Cestoda of parasitic tapeworms. The marine tapeworm Polygonoporus giganticus, found in the gut of sperm whales, can grow to over 30 m (100 ft).

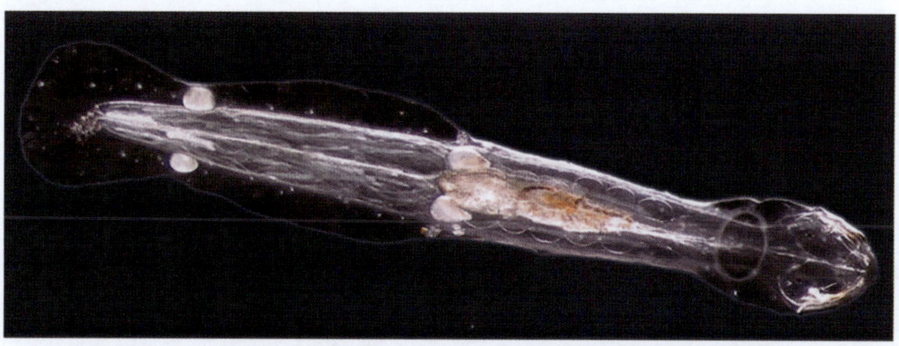

Arrow worms are predatory components of plankton worldwide.

Bloodworms are typically found on the bottom of shallow marine waters.

Bobbit worms are ambush predators that live on the seafloor.

Nematodes (roundworms) constitute a further worm phylum with tubular digestive systems and an opening at both ends. Over 25,000 nematode species have been described, of which more than half are parasitic. It has been estimated another million remain undescribed. They are ubiquitous in marine, freshwater and terrestrial environments, where they often outnumber other animals in both individual and species counts. They are found in every part of the earth's lithosphere, from the top of mountains to the bottom of oceanic trenches. By count they represent 90% of all animals on the ocean floor. Their numerical dominance, often exceeding a million individuals per square meter and accounting for about 80% of all individual animals on earth, their diversity of life cycles, and their presence at various trophic levels point at an important role in many ecosystems.

Echinoderms

Echinoderms is a phylum which contains only marine invertebrates. The adults are recognizable by their radial symmetry (usually five-point) and include starfish, sea urchins, sand dollars, and sea cucumbers, as well as the sea lilies. Echinoderms are found at every ocean depth, from the intertidal zone to the abyssal zone. The phylum contains about 7000 living species, making it the second-largest grouping of deuterostomes (a superphylum), after the chordates (which include the vertebrates, such as birds, fishes, mammals, and reptiles).

Echinoderms are unique among animals in having bilateral symmetry at the larval stage, but five-fold symmetry (pentamerism, a special type of radial symmetry) as adults.

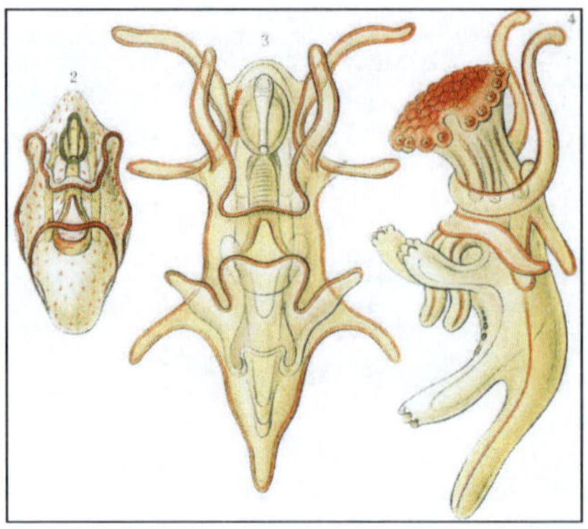

Starfish larvae are bilaterally symmetric, whereas the adults have fivefold symmetry.

Sea cucumbers filter feed on plankton and suspended solids.

The ochre sea star was the first keystone predator to be studied. They limit mussels which can overwhelm intertidal communities.

The echinoderms are important both biologically and geologically. Biologically, there are few other groupings so abundant in the biotic desert of the deep sea, as well as shallower oceans. Most echinoderms are able to regenerate tissue, organs, limbs, and reproduce asexually; in some cases, they can undergo complete regeneration from a single limb. Geologically, the value of echinoderms is in their ossified skeletons, which are major contributors to many limestone formations, and can provide valuable clues as to the geological environment. They were the most used species in regenerative research in the 19th and 20th centuries. Further, it is held by some scientists that the radiation of echinoderms was responsible for the Mesozoic Marine Revolution.

Aside from the hard-to-classify Arkarua (a Precambrian animal with echinoderm-like pentamerous radial symmetry), the first definitive members of the phylum appeared near the start of the Cambrian.

Marine Molluscs

Molluscs form a phylum with about 85,000 extant recognized species. By species count they are the largest marine phylum, comprising about 23% of all the named marine organisms. Molluscs have more varied forms than other invertebrate phylums. They are highly diverse, not just in size and in anatomical structure, but also in behavior and in habitat. The majority of species still live in the oceans, from the seashores to the abyssal zone, but some form a significant part of the freshwater fauna and the terrestrial ecosystems.

The mollusc phylum is divided into 9 or 10 taxonomic classes, two of which are extinct. These classes include gastropods, bivalves and cephalopods, as well as other lesser-known but distinctive classes. Gastropods with protective shells are referred to as snails (sea snails), whereas gastropods without protective shells are referred to as slugs (sea slugs). Gastropods are by far the most numerous molluscs in terms of classified species, accounting for 80% of the total. Bivalves include clams, oysters, cockles, mussels, scallops, and numerous other families. There are about 8,000 marine bivalves species (including brackish water and estuarine species), and about 1,200 freshwater species. Cephalopod include octopus, squid and cuttlefish. They are found in all oceans, and neurologically are the most advanced of the invertebrates. About 800 living species of marine cephalopods have been identified, and an estimated 11,000 extinct taxa have been described. There are no fully freshwater cephalopods.

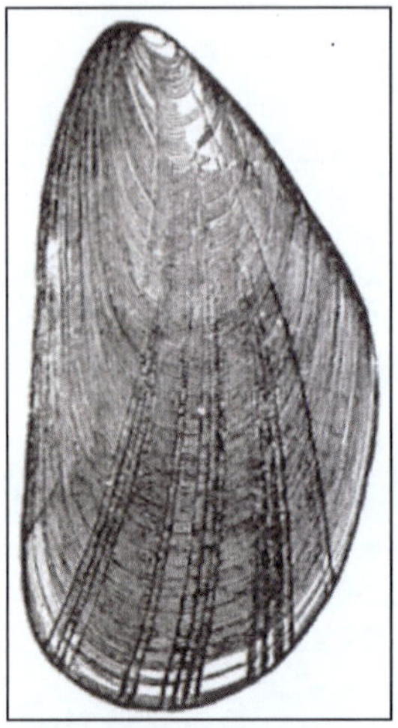

Common mussel, another bivalve.

The nautilus is a living fossil little changed since it evolved 500 million years ago as one of the first cephalopods.

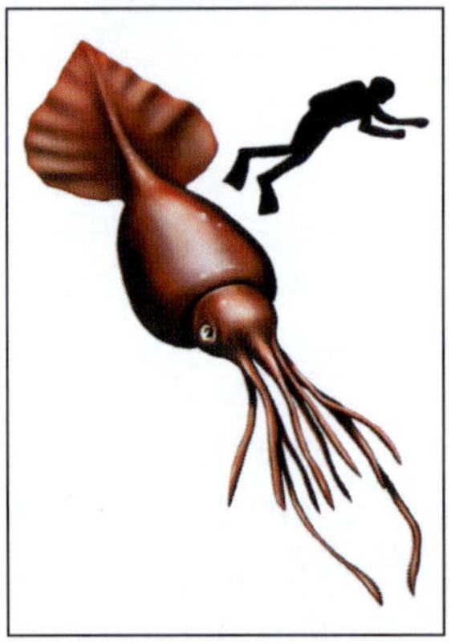

Colossal squid, largest of all invertebrates.

The sea snail Syrinx aruanus has the largest shell of any living gastropod.

Molluscs have such diverse shapes that many textbooks base their descriptions of molluscan anatomy on a generalized or hypothetical ancestral mollusc. This generalized mollusc is unsegmented and bilaterally symmetrical with an underside consisting of a single muscular foot. Beyond that it has three further key features. Firstly, it has a muscular cloak called a mantle covering its viscera and containing a significant cavity used for breathing and excretion. A shell secreted by the mantle covers the upper surface. Secondly (apart from bivalves) it has a rasping tongue called a radula used for feeding. Thirdly, it has a nervous system including a complex digestive system using microscopic, muscle-powered hairs called cilia to exude mucus. The generalized mollusc has two paired nerve cords (three in bivalves). The brain, in species that have one, encircles the esophagus.

Most molluscs have eyes and all have sensors detecting chemicals, vibrations, and touch. The simplest type of molluscan reproductive system relies on external fertilization, but more complex variations occur. All produce eggs, from which may emerge trochophore larvae, more complex veliger larvae, or miniature adults. The depiction is rather similar to modern monoplacophorans, and some suggest it may resemble very early molluscs.

Good evidence exists for the appearance of marine gastropods, cephalopods and bivalves in the Cambrian period 541 to 485.4 million years ago. However, the evolutionary history both of molluscs' emergence from the ancestral Lophotrochozoa and of their diversification into the well-known living and fossil forms are still subjects of vigorous debate among scientists.

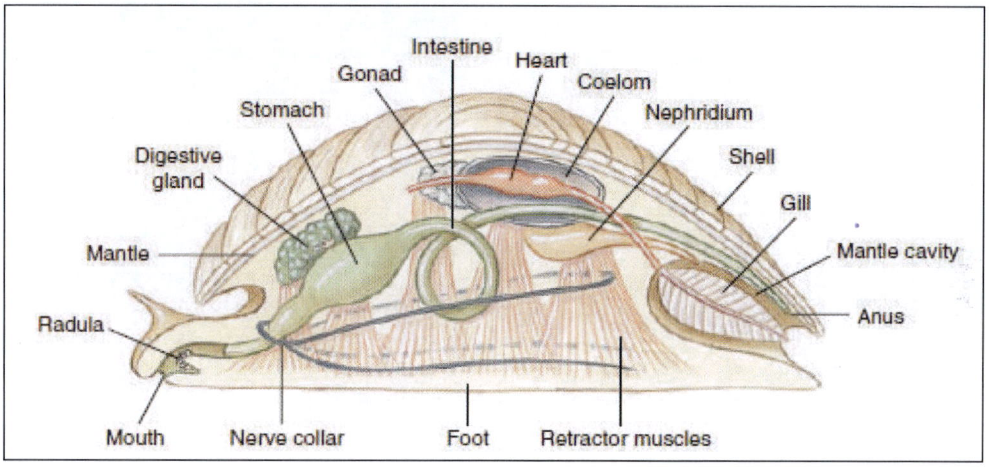

Generalized or hypothetical ancestral mollusc.

Marine Arthropods

Arthropods have an exoskeleton (external skeleton), a segmented body, and jointed appendages (paired appendages). They form a phylum which includes insects, arachnids, myriapods, and crustaceans. Arthropods are characterized by their jointed limbs and cuticle made of chitin, often mineralised with calcium carbonate. The arthropod body plan consists of segments, each with a pair of appendages. The rigid cuticle inhibits growth, so arthropods replace it periodically by moulting. Their versatility has enabled them to become the most species-rich members of all ecological guilds in most environments.

Marine arthropods range in size from the microscopic crustacean Stygotantulus to the Japanese spider crab. Arthropods' primary internal cavity is a hemocoel, which accommodates their internal organs, and through which their haemolymph - analogue of blood - circulates; they have open circulatory systems. Like their exteriors, the internal organs of arthropods are generally built of repeated segments. Their nervous system is "ladder-like", with paired ventral nerve cords running through all segments and forming paired ganglia in each segment. Their heads are formed by fusion of varying numbers of segments, and their brains are formed by fusion of the ganglia of these segments and encircle the esophagus. The respiratory and excretory systems of arthropods vary, depending as much on their environment as on the subphylum to which they belong.

Their vision relies on various combinations of compound eyes and pigment-pit ocelli: in most species the ocelli can only detect the direction from which light is coming, and the compound eyes are the main source of information, but the main eyes of spiders are ocelli that can form images and, in a few cases, can swivel to track prey. Arthropods also have a wide range of chemical and mechanical sensors, mostly based on modifications of the many setae (bristles) that project through their cuticles. Arthropods' methods of reproduction and development are diverse; all terrestrial species use internal fertilization, but this is often by indirect transfer of the sperm via an appendage or the ground, rather than by direct injection. Marine species all lay eggs and use either internal or external fertilization. Arthropod hatchlings vary from miniature adults to grubs that lack jointed limbs and eventually undergo a total metamorphosis to produce the adult form.

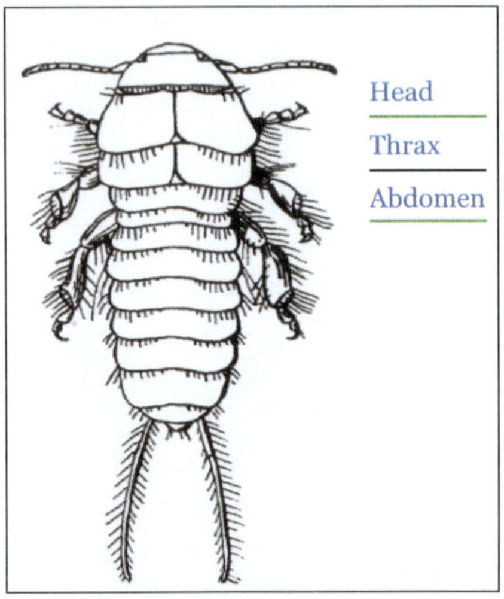

Segmentation and tagmata of an arthropod:518–522.

Horseshoe crab, a living fossil arthropod from 450 million years ago.

Crustaceans

Many crustaceans are very small, like this tiny amphipod, and make up a significant part of the ocean's zooplankton.

The banded cleaner shrimp is a crustacean common in the tropics.

The evolutionary ancestry of arthropods dates back to the Cambrian period. The group is generally regarded as monophyletic, and many analyses support the placement of arthropods with cycloneuralians (or their constituent clades) in a superphylum Ecdysozoa. Overall however, the basal relationships of Metazoa are not yet well resolved. Likewise, the relationships between various arthropod groups are still actively debated.

Other Phyla

- Tardigrade, Lobopodia, (Onychophora)
- Non-craniate (non-vertebrate) chordates: Cephalochordate, Tunicata and Haikouella. These invertebrates are close relatives of the vertebrates.

Non-Craniate Chordates are Close Relatives of Vertebrates

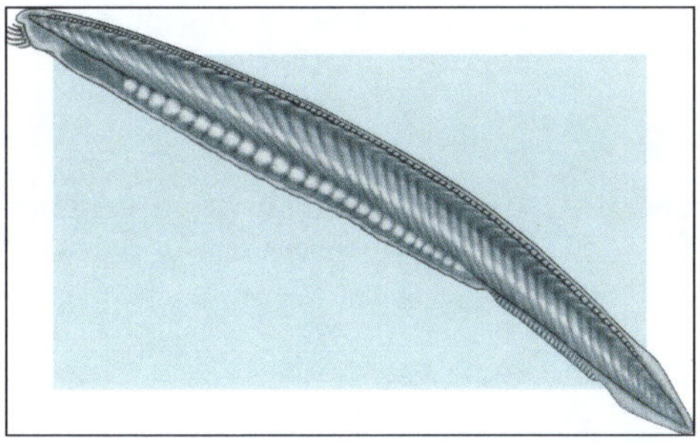

The lancelet, a small translucent fish-like Cephalochordate, is the closest living invertebrate relative of the vertebrates.

Fluorescent-colored sea squirts, Rhopalaea crassa. Tunicates may provide clues to vertebrate (and therefore human) ancestry.

Salp chain.

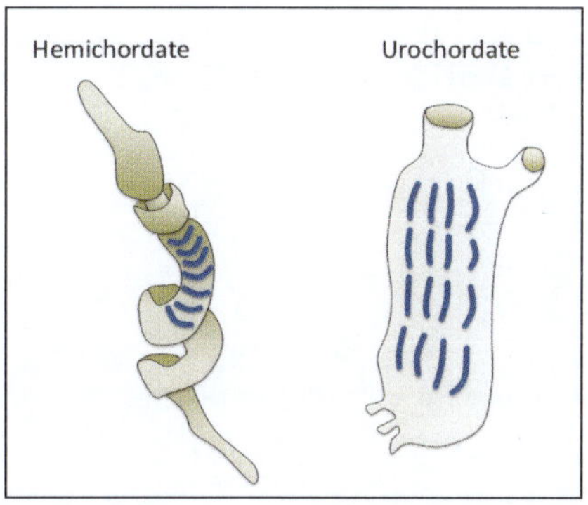

Gill slits in an acorn worm (left) and tunicate (right).

Minerals from Sea Water

There are a number of marine invertebrates that use minerals that are present in the sea in such minute quantities that they were undetectable until the advent of spectroscopy. Vanadium is concentrated by some tunicates for use in their blood cells to a level ten million times that of the surrounding seawater. Other tunicates similarly concentrate niobium and tantalum. Lobsters use copper in their respiratory pigment hemocyanin, despite the proportion of this metal in seawater being minute. Although these elements are present in vast quantities in the ocean, their extraction by man is not economic.

References

- Pace NR (May 2006). "Time for a change". Nature. 441(7091): 289. Bibcode:2006Natur.441..289P. Doi:10.1038/441289a. PMID 16710401
- Marine-life-definition-and-examples: thoughtco.com, Retrieved 21 May, 2019
- Borenstein, Seth (19 October 2015). "Hints of life on what was thought to be desolate early Earth". Associated Press. Retrieved 9 October 2018
- Jones, E.B.G.; Pang, K.-L., eds. (2012). Marine Fungi, and Fungal-like Organisms. Marine and Freshwater Botany. Berlin, Boston: De Gruyter (published August 2012). Doi:10.1515/9783110264067. ISBN 978-3-11-026406-7. Retrieved 3 September 2015
- Marine-gardens-5-types-plants-ocean, explore-the-blue: leisurepro.com, Retrieved 22 June, 2019
- De La Rocha, C.L.; Passow, U. (2014). Treatise on Geochemistry. Pp. 93–122. Doi:10.1016/b978-0-08-095975-7.00604-5. ISBN 9780080983004
- Seagrass-and-seagrass-beds, plants-algae, ocean-life: ocean-life, Retrieved 23 July, 2019
- Jude, B.A., Kirn, T.J., Taylor R.K. (2005). "A colonization factor links Vibrio cholerae environmental survival and human infection". Nature. 438 (7069): 863–6. Doi:10.1038/nature04249. PMID 16341015
- Guideline-Marine-Fauna-131216-2, Policies-and-Guidance: epa.wa.gov.au, Retrieved 24 August, 2019
- Peschek, Günter A.; Bernroitner, Margit; Sari, Samira; Pairer, Martin; Obinger, Christian (2011). Bioenergetic Processes of Cyanobacteria. Springer, Dordrecht. Pp. 3–70. Doi:10.1007/978-94-007-0388-9_1. ISBN 9789400703520
- Phytoplankton, features: nasa.gov, Retrieved 25 January, 2019
- Bobrovskiy, Ilya (2018). "Ancient steroids establish the Ediacaran fossil Dickinsonia as one of the earliest animals". Science. 361 (6408): 1246-1249. doi:10.1126/science.aat7228. PMID 30237355
- Cool-types-marine-worms, explore-the-blue: leisurepro.com, Retrieved 27 March 2019
- Kennish, Michael J. (2001). Practical handbook of marine science. Marine science series (3rd ed.). CRC Press. P. 35. ISBN 978-0-8493-2391-1

Impact of Human Activities on Marine Ecology

4

- **Fishery**
- **Marine Pollution**
- **Deep Sea Mining**
- **Marine Debris**
- **Deepwater Drilling**
- **Effects of Oil Drilling on the Ocean**

There are numerous human activities which have an adverse effect upon marine ecology. A few of these are deep sea mining, deepwater drilling and fish farming. They can have numerous consequences such as marine pollution and accumulation of marine debris. This chapter discusses in detail these human activities that negatively affect the marine ecology.

Fishery

A fishery (plural: fisheries) is an organized effort by humans to catch fish or other aquatic species, an activity known as fishing.

Generally, a fishery exists for the purpose of providing human food, although other aims are possible (such as sport or recreational fishing), or obtaining ornamental fish or fish products such as fish oil.

Regardless of purpose, however, the term fishery generally refers to a fishing effort centered on either a particular ecoregion or a particular species or type of fish or aquatic animal, and usually fisheries are differentiated by both criteria.

Examples would be the salmon fishery of Alaska, the cod fishery off the Lofoten islands or the tuna fishery of the Eastern Pacific.

Most fisheries are marine, rather than freshwater; most marine fisheries are based near the coast.

This is not only because harvesting from relatively shallow waters is easier than in the open ocean, but also because fish are much more abundant near the coastal shelf, due to coastal upwelling and the abundance of nutrients available there.

Impact of Fish Farming

The most common negative environmental impacts that have been associated with aquaculture include: waters eutrophication, water quality, alteration or destruction of natural habitats; introduction and transmission of aquatic animal diseases.

Fish-farming contributing to marine waters eutrophication, faeces and uneaten food pellets from fish farms alter the organic matter in the sediment, which can change the consumption of oxygen and cause local eutrophication. On 2005 Zaccone et al. investigated on fish-farming impact on both pelagic and sediment environments of Mar Piccolo of Taranto (Italy), studying the distribution of microbiological indicators (heterotrophic bacteria and coliforms), in relation to physical and chemical parameters showing that the benthonic environment increased the bio-deposition of the sediment, causing changes in the abundance and the composition (heterotrophic bacteria/clostridia ratio) of microflora. New methods were developed for the detection of environmental indicators. In addition to the eutrofication that cause huge damages for the environment and local fauna, the spread and the outbreaks of diseases are negative consequences of the expansion and diversification of the aquaculture sector associated with globalization. The development and implementation of national aquatic animal health management plans is important to prevent, control and eliminate diseases in a timely manner and respond to consumers increasing concerns for food safety, ecosystems integrity and animal welfare.

How to Prevent the Impacts and to Improve the Quality Product?

Aquaculture needs an enabling policy environment in order to grow in a sustainable manner and to be integrated into the coastal zone. Moreover the interactions between aquaculture and the larger system in which it occurs, in particular, the influence of the surrounding natural and social environment on aquaculture, must be taken into consideration.

In my opinion only an integrated and multitasks vision can improve the aquaculture industry, in fact it's very important take account of the full range of ecosystem functions and services and should not threaten the sustained delivery of these to society.

Marine Pollution

Marine pollution occurs when harmful effects result from the entry into the ocean of chemicals, particles, industrial, agricultural, and residential waste, noise, or the spread of invasive organisms. Eighty percent of marine pollution comes from land. Air pollution is also a contributing factor by carrying off pesticides or dirt into the ocean. Land and air pollution have proven to be harmful to marine life and its habitats.

The pollution often comes from nonpoint sources such as agricultural runoff, wind-blown debris, and dust. Nutrient pollution, a form of water pollution, refers to contamination by excessive inputs of nutrients. It is a primary cause of eutrophication of surface waters, in which excess nutrients, usually nitrates or phosphates, stimulate algae growth. Many potentially toxic chemicals adhere to tiny particles which are then taken up by plankton and benthic animals, most of which are either deposit feeders or filter feeders. In this way, the toxins are concentrated upward within ocean food chains. Many particles combine chemically in a manner highly depletive of oxygen, causing estuaries to become anoxic.

When pesticides are incorporated into the marine ecosystem, they quickly become absorbed into marine food webs. Once in the food webs, these pesticides can cause mutations, as well as diseases, which can be harmful to humans as well as the entire food web. Toxic metals can also be introduced into marine food webs. These can cause a change to tissue matter, biochemistry, behavior, reproduction, and suppress growth in marine life. Also, many animal feeds have a high fish meal or fish hydrolysate content. In this way, marine toxins can be transferred to land animals, and appear later in meat and dairy products.

In order to protect the ocean from marine pollution, policies have been developed internationally. There are different ways for the ocean to get polluted, therefore there have been multiple laws, policies, and treaties put into place throughout history.

While marine pollution can be obvious, as with the marine debris shown above, it is often the pollutants that cannot be seen that cause most harm.

Pathways of Pollution

There are many ways to categorize and examine the inputs of pollution into our marine ecosystems. Patin (n.d.) notes that generally there are three main types of inputs of pollution into the ocean: direct discharge of waste into the oceans, runoff into the waters due to rain, and pollutants released from the atmosphere.

One common path of entry by contaminants to the sea are rivers. The evaporation of water from oceans exceeds precipitation. The balance is restored by rain over the continents entering rivers and then being returned to the sea. The Hudson in New York State and the Raritan in New Jersey, which empty at the northern and southern ends of Staten Island, are a source of mercury contamination of zooplankton (copepods) in the open ocean. The highest concentration in the filter-feeding copepods is not at the mouths of these rivers but 70 miles (110 km) south, nearer Atlantic City, because water flows close to the coast. It takes a few days before toxins are taken up by the plankton.

Pollution is often classed as point source or nonpoint source pollution. Point source pollution occurs when there is a single, identifiable, localized source of the pollution. An example is directly discharging sewage and industrial waste into the ocean. Pollution such as this occurs particularly in developing nations. Nonpoint source pollution occurs when the pollution comes from ill-defined and diffuse sources. These can be difficult to regulate. Agricultural runoff and wind blown debris are prime examples.

Septic river.

Direct Discharge

Pollutants enter rivers and the sea directly from urban sewerage and industrial waste discharges, sometimes in the form of hazardous and toxic wastes.

Inland mining for copper, gold, etc., is another source of marine pollution. Most of the pollution is simply soil, which ends up in rivers flowing to the sea. However, some minerals discharged in the course of the mining can cause problems, such as copper, a common industrial pollutant, which can interfere with the life history and development of coral polyps. Mining has a poor environmental track record. For example, according to the United States Environmental Protection Agency, mining has contaminated portions of the headwaters of over 40% of watersheds in the western continental US. Much of this pollution finishes up in the sea.

Acid mine drainage in the Rio Tinto River.

Land Runoff

Surface runoff from farming, as well as urban runoff and runoff from the construction of roads, buildings, ports, channels, and harbours, can carry soil and particles laden with carbon, nitrogen, phosphorus, and minerals. This nutrient-rich water can cause fleshy algae and phytoplankton to thrive in coastal areas; known as algal blooms, which have the potential to create hypoxic conditions by using all available oxygen. In the coast of southwest Florida, harmful algal blooms have existed for over 100 years. These algal blooms have been a cause of species of fish, turtles, dolphins, and shrimp to die and cause harmful effects on humans who swim in the water.

Polluted runoff from roads and highways can be a significant source of water pollution in coastal areas. About 75% of the toxic chemicals that flow into Puget Sound are carried by stormwater that runs off paved roads and driveways, rooftops, yards and other developed land. In California, there are many rainstorms that runoff into the ocean. These rainstorms occur from October to March, and these runoff waters contain petroleum, heavy metals, pollutants from emissions, etc.

In China, there is a large coastal population that pollutes the ocean through land runoff. This includes sewage discharge and pollution from urbanization and land use. In 2001, more than 66,795 mi^2 of the Chinese coastal ocean waters were rated less than Class I of the Sea Water Quality Standard of China. Much of this pollution came from Ag, Cu, Cd, Pb, As, DDT, PCBs, etc., which occurred from contamination through land runoff.

Ship Pollution

Ships can pollute waterways and oceans in many ways. Oil spills can have devastating effects. While being toxic to marine life, polycyclic aromatic hydrocarbons (PAHs), found in crude oil, are very difficult to clean up, and last for years in the sediment and marine environment.

Oil spills are probably the most emotive of marine pollution events. However, while a tanker wreck may result in extensive newspaper headlines, much of the oil in the world's seas comes from other smaller sources, such as tankers discharging ballast water from oil tanks used on return ships, leaking pipelines or engine oil disposed of down sewers.

Discharge of cargo residues from bulk carriers can pollute ports, waterways, and oceans. In many instances vessels intentionally discharge illegal wastes despite foreign and domestic regulation prohibiting such actions. An absence of national standards provides an incentive for some cruise liners to dump waste in places where the penalties are inadequate. It has been estimated that container ships lose over 10,000 containers at sea each year (usually during storms). Ships also create noise pollution that disturbs natural wildlife, and water from ballast tanks can spread harmful algae and other invasive species.

Ballast water taken up at sea and released in port is a major source of unwanted exotic marine life. The invasive freshwater zebra mussels, native to the Black, Caspian, and Azov seas, were probably transported to the Great Lakes via ballast water from a transoceanic vessel. Meinesz believes that one of the worst cases of a single invasive species causing harm to an ecosystem can be attributed to a seemingly harmless jellyfish. Mnemiopsis leidyi, a species of comb jellyfish that spread so it now inhabits estuaries in many parts of the world. It was first introduced in 1982, and thought to have been transported to the Black Sea in a ship's ballast water. The population of the jellyfish grew exponentially and, by 1988, it was wreaking havoc upon the local fishing industry. "The anchovy catch fell from 204,000 tons in 1984 to 200 tons in 1993; sprat from 24,600 tons in 1984 to 12,000 tons in 1993; horse mackerel from 4,000 tons in 1984 to zero in 1993." Now that the jellyfish have exhausted the zooplankton, including fish larvae, their numbers have fallen dramatically, yet they continue to maintain a stranglehold on the ecosystem.

Invasive species can take over once occupied areas, facilitate the spread of new diseases, introduce new genetic material, alter underwater seascapes, and jeopardize the ability of native species to obtain food. Invasive species are responsible for about $138 billion annually in lost revenue and management costs in the US alone.

A cargo ship pumps ballast water over the side.

Atmospheric Pollution

Another pathway of pollution occurs through the atmosphere. Wind-blown dust and debris, including plastic bags, are blown seaward from landfills and other areas. Dust from the Sahara moving around the southern periphery of the subtropical ridge moves into the Caribbean and Florida during the warm season as the ridge builds and moves northward through the subtropical Atlantic. Dust can also be attributed to a global transport from the Gobi and Taklamakan deserts across Korea, Japan, and the Northern Pacific to the Hawaiian Islands.

Since 1970, dust outbreaks have worsened due to periods of drought in Africa. There is a large variability in dust transport to the Caribbean and Florida from year to year; however, the flux is greater during positive phases of the North Atlantic Oscillation. The USGS links dust events to a decline in the health of coral reefs across the Caribbean and Florida, primarily since the 1970s.

Climate change is raising ocean temperatures and raising levels of carbon dioxide in the atmosphere. These rising levels of carbon dioxide are acidifying the oceans. This, in turn, is altering aquatic ecosystems and modifying fish distributions, with impacts on the sustainability of fisheries and the livelihoods of the communities that depend on them. Healthy ocean ecosystems are also important for the mitigation of climate change.

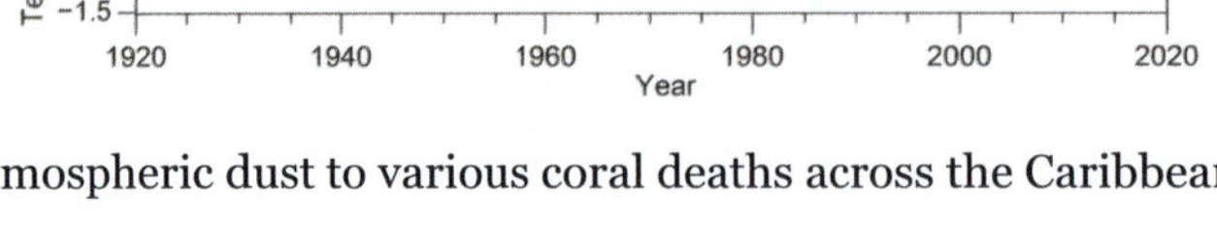

A graph linking atmospheric dust to various coral deaths across the Caribbean Sea and Florida.

Deep Sea Mining

Deep sea mining is a relatively new mineral retrieval process that takes place on the ocean floor. Ocean mining sites are usually around large areas of polymetallic nodules or active and extinct hydrothermal vents at about 1,400 – 3,700 meters below the ocean's surface. The vents create sulfide deposits, which contain precious metals such as silver, gold, copper, manganese, cobalt, and zinc. The deposits are mined using either hydraulic pumps or bucket systems that take ore to the surface to be processed. As with all mining operations, deep sea mining raises questions about environmental damages to the surrounding areas.

Because deep sea mining is a relatively new field, the complete consequences of full-scale mining operations are unknown. However, experts are certain that removal of parts of the sea floor will result in disturbances to the benthic layer, increased toxicity of the water column, and sediment plumes from tailings. Removing parts of the sea floor disturbs the habitat of benthic organisms, possibly, depending on the type of mining and location, causing permanent disturbances. Aside from direct impact of mining the area, leakage, spills, and corrosion would alter the mining area's chemical makeup.

Among the impacts of deep sea mining, sediment plumes could have the greatest impact. Plumes are caused when the tailings from mining (usually fine particles) are dumped back into the ocean, creating a cloud of particles floating in the water. Two types of plumes occur: near-bottom plumes and surface plumes. Near-bottom plumes occur when the tailings are pumped back down to the mining site. The floating particles increase the turbidity, or cloudiness, of the water, clogging filter-feeding apparatuses used by benthic organisms. Surface plumes cause a more serious problem. Depending on the size of the particles and water currents the plumes could spread over vast areas. The plumes could impact zooplankton and light penetration, in turn affecting the food web of the area.

Types of Pollution

Acidification

The oceans are normally a natural carbon sink, absorbing carbon dioxide from the atmosphere. Because the levels of atmospheric carbon dioxide are increasing, the oceans are becoming more acidic. The potential consequences of ocean acidification are not fully understood, but there are concerns that structures made of calcium carbonate may become vulnerable to dissolution, affecting corals and the ability of shellfish to form shells.

Oceans and coastal ecosystems play an important role in the global carbon cycle and have removed about 25% of the carbon dioxide emitted by human activities between 2000 and 2007 and about half the anthropogenic CO_2 released since the start of the industrial revolution. Rising ocean temperatures and ocean acidification means that the capacity of the ocean carbon sink will gradually get weaker, giving rise to global concerns expressed in the Monaco and Manado Declarations.

A report from NOAA scientists published in the journal Science in May 2008 found that large amounts of relatively acidified water are upwelling to within four miles of the Pacific continental shelf area of North America. This area is a critical zone where most local marine life lives or is born. While the paper dealt only with areas from Vancouver to northern California, other continental shelf areas may be experiencing similar effects.

A related issue is the methane clathrate reservoirs found under sediments on the ocean floors. These trap large amounts of the greenhouse gas methane, which ocean warming has the potential to release. In 2004 the global inventory of ocean methane clathrates was estimated to occupy between one and five million cubic kilometres. If all these clathrates were to be spread uniformly across the ocean floor, this would translate to a thickness between three and fourteen metres. This estimate corresponds to 500–2500 gigatonnes carbon (Gt C), and can be compared with the 5000 Gt C estimated for all other fossil fuel reserves.

An island with a fringing reef in the Maldives. Coral reefs are dying around the world.

Eutrophication

Eutrophication is an increase in chemical nutrients, typically compounds containing nitrogen or phosphorus, in an ecosystem. It can result in an increase in the ecosystem's primary productivity (excessive plant growth and decay), and further effects including lack of oxygen and severe reductions in water quality, fish, and other animal populations.

The biggest culprit are rivers that empty into the ocean, and with it the many chemicals used as fertilizers in agriculture as well as waste from livestock and humans. An excess of oxygen-depleting chemicals in the water can lead to hypoxia and the creation of a dead zone.

Estuaries tend to be naturally eutrophic because land-derived nutrients are concentrated where runoff enters the marine environment in a confined channel. The World Resources Institute has identified 375 hypoxic coastal zones around the world, concentrated in coastal areas in Western Europe, the Eastern and Southern coasts of the US, and East Asia, particularly in Japan. In the ocean, there are frequent red tide algae blooms that kill fish and marine mammals and cause respiratory problems in humans and some domestic animals when the blooms reach close to shore.

In addition to land runoff, atmospheric anthropogenic fixed nitrogen can enter the open ocean. A study in 2008 found that this could account for around one third of the ocean's external (non-recycled) nitrogen supply and up to three per cent of the annual new marine biological production. It has been suggested that accumulating reactive nitrogen in the environment may have consequences as serious as putting carbon dioxide in the atmosphere.

One proposed solution to eutrophication in estuaries is to restore shellfish populations, such as oysters. Oyster reefs remove nitrogen from the water column and filter out suspended solids, subsequently reducing the likelihood or extent of harmful algal blooms or anoxic conditions. Filter feeding activity is considered beneficial to water quality by controlling phytoplankton density and sequestering nutrients, which can be removed from the system through shellfish harvest, buried in the sediments, or lost through denitrification. Foundational work toward the idea of improving marine water quality through shellfish cultivation to was conducted by Odd Lindahl et al., using mussels in Sweden.

A polluted lagoon.

The effect of eutrophication on marine benthic life.

Plastic Debris

Marine debris is mainly discarded human rubbish which floats on, or is suspended in the ocean. Eighty percent of marine debris is plastic – a component that has been rapidly accumulating since the end of World War II. The mass of plastic in the oceans may be as high as 100,000,000 tonnes (98,000,000 long tons; 110,000,000 short tons).

Discarded plastic bags, six pack rings, cigarette butts and other forms of plastic waste which finish up in the ocean present dangers to wildlife and fisheries. Aquatic life can be threatened through entanglement, suffocation, and ingestion. Fishing nets, usually made of plastic, can be left or lost in the ocean by fishermen. Known as ghost nets, these entangle fish, dolphins, sea turtles, sharks, dugongs, crocodiles, seabirds, crabs, and other creatures, restricting movement, causing starvation, laceration, infection, and, in those that need to return to the surface to breathe, suffocation.

Many animals that live on or in the sea consume flotsam by mistake, as it often looks similar to their natural prey. Plastic debris, when bulky or tangled, is difficult to pass, and may become permanently lodged in the digestive tracts of these animals. Especially when evolutionary adaptions make it impossible for the likes of turtles to reject plastic bags, which resemble jellyfish when immersed in water, as they have a system in their throat to stop slippery foods from otherwise escaping. Thereby blocking the passage of food and causing death through starvation or infection.

Plastics accumulate because they don't biodegrade in the way many other substances do. They will photodegrade on exposure to the sun, but they do so properly only under dry conditions, and water inhibits this process. In marine environments, photodegraded plastic disintegrates into ever-smaller pieces while remaining polymers, even down to the molecular level. When floating plastic particles photodegrade down to zooplankton sizes, jellyfish attempt to consume them, and in this way the plastic enters the ocean food chain.

Many of these long-lasting pieces end up in the stomachs of marine birds and animals, including sea turtles, and black-footed albatross. In a 2008 Pacific Gyre voyage, Algalita Marine Research Foundation researchers began finding that fish are ingesting plastic fragments and debris. Of the 672 fish caught during that voyage, 35% had ingested plastic pieces.

Plastic debris tends to accumulate at the centre of ocean gyres. The North Pacific Gyre, for example, has collected the so-called "Great Pacific Garbage Patch", which is now estimated to be one to twenty times the size of Texas (approximately from 700,000 to 15,000,000 square kilometers). There could be as much plastic as fish in the sea. It has a very high level of plastic particulate suspended in the upper water column. In samples taken in 1999, the mass of plastic exceeded that of zooplankton (the dominant animal life in the area) by a factor of six.

Midway Atoll, in common with all the Hawaiian Islands, receives substantial amounts of debris from the garbage patch. Ninety percent plastic, this debris accumulates on the beaches of Midway where it becomes a hazard to the bird population of the island. Midway Atoll is home to two-thirds (1.5 million) of the global population of Laysan albatross. Nearly all of these albatross have plastic in their digestive system and one-third of their chicks die.

Toxic additives used in the manufacture of plastic materials can leach out into their surroundings when exposed to water. Waterborne hydrophobic pollutants collect and magnify on the surface of plastic debris, thus making plastic far more deadly in the ocean than it would be on land. Hydrophobic contaminants are also known to bioaccumulate in fatty tissues, biomagnifying up the food chain and putting pressure on apex predators. Some plastic additives are known to disrupt the endocrine system when consumed, others can suppress the immune system or decrease reproductive rates.

Floating debris can also absorb persistent organic pollutants from seawater, including PCBs, DDT, and PAHs. Aside from toxic effects, when ingested some of these affect animal brain cells similarly to estradiol, causing hormone disruption in the affected wildlife. Saido, a chemist with the College of Pharmacy, conducted a study in Nihon University, Chiba, Japan, that discovered, when plastics eventually decompose, they produce potentially toxic bisphenol A (BPA) and PS oligomer into the water. These toxins are believed to bring harm to the marine life living in the area.

A mute swan builds a nest using plastic garbage.

A growing concern regarding plastic pollution in the marine ecosystem is the use of microplastics. Microplastics are little beads of plastic less than 5 millimeters wide, and they are commonly found in hand soaps, face cleansers, and other exfoliators. When these products are used, the microplastics go through the water filtration system and into the ocean, but because of their small size they are likely to escape capture by the preliminary treatment screens on wastewater plants. These beads are harmful to the organisms in the ocean, especially filter feeders, because they can easily ingest the plastic and become sick. The microplastics are such a concern because it is difficult to clean them up due to their size, so humans can try to avoid using these harmful plastics by purchasing products that use environmentally safe exfoliates.

The remains of an albatross containing ingested flotsam.

Toxins

Apart from plastics, there are particular problems with other toxins that do not disintegrate rapidly in the marine environment. Examples of persistent toxins are PCBs, DDT, TBT, pesticides, furans, dioxins, phenols, and radioactive waste. Heavy metals are metallic chemical elements that have a relatively high density and are toxic or poisonous at low concentrations. Examples are mercury, lead, nickel, arsenic, and cadmium. Such toxins can accumulate in the tissues of many species of aquatic life in a process called bioaccumulation. They are also known to accumulate in benthic environments, such as estuaries and bay muds: a geological record of human activities of the last century.

Specific Examples

- Chinese and Russian industrial pollution such as phenols and heavy metals in the Amur River have devastated fish stocks and damaged its estuary soil.
- Wabamun Lake in Alberta, Canada, once the best whitefish lake in the area, now has unacceptable levels of heavy metals in its sediment and fish.
- Acute and chronic pollution events have been shown to impact southern California kelp forests, though the intensity of the impact seems to depend on both the nature of the contaminants and duration of exposure.
- Due to their high position in the food chain and the subsequent accumulation of heavy metals from their diet, mercury levels can be high in larger species such as bluefin and albacore. As a result, in March 2004 the United States FDA issued guidelines recommending that pregnant women, nursing mothers and children limit their intake of tuna and other types of predatory fish.
- Some shellfish and crabs can survive polluted environments, accumulating heavy metals or toxins in their tissues. For example, mitten crabs have a remarkable ability to survive in

highly modified aquatic habitats, including polluted waters. The farming and harvesting of such species needs careful management if they are to be used as a food.

- Surface runoff of pesticides can alter the gender of fish species genetically, transforming male into female fish.
- Heavy metals enter the environment through oil spills – such as the Prestige oil spill on the Galician coast and Gulf of Mexico which unleashed an estimated 3.19 million barrels of oil – or from other natural or anthropogenic sources.
- In 2005, the 'Ndrangheta, an Italian mafia syndicate, was accused of sinking at least 30 ships loaded with toxic waste, much of it radioactive. This has led to widespread investigations into radioactive waste disposal rackets.
- Since the end of World War II, various nations, including the Soviet Union, the United Kingdom, the United States, and Germany, have disposed of chemical weapons in the Baltic Sea, raising concerns of environmental contamination.
- The damaging of the Fukushima Dai-ichi Nuclear Power Plant in 2011 caused radioactive toxins to leak into the air and ocean. There are still many isotopes in the ocean, which directly affects the benthic food web and also affects the whole food chain. The concentration of 137Cs in the bottom sediment that was contaminated by water with high concentrations in April–May 2011 remains quite high and is showing signs of very slow decrease with time.

Underwater Noise

Marine life can be susceptible to noise or the sound pollution from sources such as passing ships, oil exploration seismic surveys, and naval low-frequency active sonar. Sound travels more rapidly and over larger distances in the sea than in the atmosphere. Marine animals, such as cetaceans, often have weak eyesight, and live in a world largely defined by acoustic information. This applies also to many deeper sea fish, who live in a world of darkness. Between 1950 and 1975, ambient noise at one location in the Pacific Ocean increased by about ten decibels (that is a tenfold increase in intensity).

Noise also makes species communicate louder, which is called the Lombard vocal response. Whale songs are longer when submarine-detectors are on. If creatures don't "speak" loud enough, their voice can be masked by anthropogenic sounds. These unheard voices might be warnings, finding of prey, or preparations of net-bubbling. When one species begins speaking louder, it will mask other species voices, causing the whole ecosystem to eventually speak louder.

According to the oceanographer Sylvia Earle, "Undersea noise pollution is like the death of a thousand cuts. Each sound in itself may not be a matter of critical concern, but taken all together, the noise from shipping, seismic surveys, and military activity is creating a totally different environment than existed even 50 years ago. That high level of noise is bound to have a hard, sweeping impact on life in the sea."

Noise from ships and human activity can damage Cnidarians and Ctenophora, which are very important organisms in the marine ecosystem. They promote high diversity and they are used as models for ecology and biology because of their simple structures. When there is underwater noise, the vibrations in the water damage the cilia hairs in the Coelenterates. In a study, the organisms were exposed to sound waves for different numbers of times and the results showed that damaged hair cells were extruded or missing or presented bent, flaccid or missed kinocilia and stereocilia.

Adaptation and Mitigation

Much anthropogenic pollution ends up in the ocean. The 2011 edition of the United Nations Environment Programme Year Book identifies as the main emerging environmental issues the loss to the oceans of massive amounts of phosphorus, "a valuable fertilizer needed to feed a growing global population", and the impact billions of pieces of plastic waste are having globally on the health of marine environments.

Bjorn Jennssen notes in his article, "Anthropogenic pollution may reduce biodiversity and productivity of marine ecosystems, resulting in reduction and depletion of human marine food resources". There are two ways the overall level of this pollution can be mitigated: either the human population is reduced, or a way is found to reduce the ecological footprint left behind by the average human. If the second way is not adopted, then the first way may be imposed as the world ecosystems falter.

The second way is for humans, individually, to pollute less. That requires social and political will, together with a shift in awareness so more people respect the environment and are less disposed to abuse it. At an operational level, regulations, and international government participation is needed. It is often very difficult to regulate marine pollution because pollution spreads over international barriers, thus making regulations hard to create as well as enforce.

Without appropriate awareness of marine pollution, the necessary global will to effectively address the issues may prove inadequate. Balanced information on the sources and harmful effects of marine pollution need to become part of general public awareness, and ongoing research is required to fully establish, and keep current, the scope of the issues. As expressed in Daoji and Dag's research, one of the reasons why environmental concern is lacking among the Chinese is because the public awareness is low and therefore should be targeted.

The amount of awareness on marine pollution is vital to the support of keeping the prevention of trash from entering waterways and ending up in our oceans. The EPA reports that in 2014 Americans generated about 258 million tons of waste, and only a third was recycled or composted. In 2015, there was over 8 million tons of plastic that made it into the ocean. Through more sustainable packing this could lead to; eliminating toxic constituents, using fewer materials, making more readily available recyclable plastic. However, awareness can only take these initiatives so far. The most abundant plastic is PET (Polyethylene terephthalate) and is the most resistant to biodegradables. Researchers have been making great strides in combating this problem. In one way has been by adding a special polymer called a tetrablock copolymer. The tetrablock copolymer acts as a laminate between the PE and iPP which enables for an easier breakdown but still be tough. Through more awareness, individuals will become more cognizant of their carbon footprints. Also, from research and technology, more strides can be made to aid in the plastic pollution problem.

In 2018 a survey of Global Oceanic Environmental Survey (GOES) Foundation find that the ecosystem in seas and oceans can collapse in the next 25 years what would fail the terrestrial ecosystem and end the life on earth as we know them. The main causes: Plastic pollution, Ocean acidification, Ocean pollution. For prevent that scenario from happen, we need a total single use plastic ban, ban on wood burning, planting as many trees as possible, "pollution-free recycling of electronics and by 2030 all industries to be zero toxic discharge." "special protection and perservation of peat bogs, wetlands, marshlands and mangrove swamps to ensure carbon dioxide is absorbed from the atmosphere."

Aerosol can pollute beaches.

Deep Sea Mining

Deep sea mining is a mineral retrieval process that takes place on the ocean floor. Ocean mining sites are usually around large areas of polymetallic nodules or active and extinct hydrothermal vents at 1,400 to 3,700 metres (4,600 to 12,100 ft) below the ocean's surface. The vents create globular or massive sulfide deposits, which contain valuable metals such as silver, gold, copper, manganese, cobalt, and zinc. The deposits are mined using either hydraulic pumps or bucket systems that take ore to the surface to be processed. As with all mining operations, deep sea mining raises questions about its potential environmental impact. Environmental advocacy groups such as Greenpeace and the Deep sea Mining Campaign have argued that seabed mining should not be permitted in most of the world's oceans because of the potential for damage to deepsea ecosystems and pollution by heavy metal laden plumes.

The international law–based regulations on deep sea mining are contained in the United Nations Conventions on the Law of the Sea from 1973 to 1982, which came into force in 1994. The convention set up the International Seabed Authority (ISA), which regulates nations' deep sea mining ventures outside each nations' Exclusive Economic Zone (a 200-nautical-mile (370 km) area surrounding coastal nations). The ISA requires nations interested in mining to explore two equal mining sites and turn one over to the ISA, along with a transfer of mining technology over a 10- to 20-year period. This seemed reasonable at the time because it was widely believed that nodule mining would be extremely profitable. However, these strict requirements led some industrialized countries to refuse to sign the initial treaty in 1982.

The US abides by the Deep Seabed Hard Mineral Resources Act, which was originally written in 1980. This legislations is largely recognized as one of the main concerns the US has with ratifying UNCLOS.

Within the EEZ of nation states seabed mining comes under the jurisdiction of national laws. Despite extensive exploration both within and outside of EEZs, only a few countries, notably New Zealand, have established legal and institutional frameworks for the future development of deep seabed mining.

Papua New Guinea was the first country to approve a permit for the exploration of minerals in the deep seabed. Solwara 1 was awarded its licence and environmental permits despite three independent reviews of the environmental impact statement mine finding significant gaps and flaws in the underlying science.

The ISA has recently arranged a workshop in Australia where scientific experts, industry representatives, legal specialists and academics worked towards improving existing regulations and ensuring that development of seabed minerals does not cause serious and permanent damage to the marine environment.

Resources Mined

The deep sea contains many different resources available for extraction, including silver, gold, copper, manganese, cobalt, and zinc. These raw materials are found in various forms on the sea floor.

Minerals and Related Depths

Type of mineral deposit	Averge Depth	Resources found
Polymetalic nodules	4,000 – 6,000 m	Nickel, copper cobalt and manganese
Manganese crusts	800 – 2,400 m	Mainly cobalt, some vanadium, molybdenum and platinum
Sulfide deposits	1,400 – 3,700 m	Copper, lead and zinc some gold and silver

Diamonds are also mined from the seabed by De Beers and others. Nautilus Minerals Inc and Neptune Minerals are planning to mine the offshore waters of Papua New Guinea and New Zealand.

Extraction Methods

Recent technological advancements have given rise to the use remotely operated vehicles (ROVs) to collect mineral samples from prospective mine sites. Using drills and other cutting tools, the

ROVs obtain samples to be analyzed for precious materials. Once a site has been located, a mining ship or station is set up to mine the area.

There are two predominant forms of mineral extraction being considered for full-scale operations: continuous-line bucket system (CLB) and the hydraulic suction system. The CLB system is the preferred method of nodule collection. It operates much like a conveyor-belt, running from the sea floor to the surface of the ocean where a ship or mining platform extracts the desired minerals, and returns the tailings to the ocean. Hydraulic suction mining lowers a pipe to the seafloor which transfers nodules up to the mining ship. Another pipe from the ship to the seafloor returns the tailings to the area of the mining site.

In recent years, the most promising mining areas have been the Central and Eastern Manus Basin around Papua New Guinea and the crater of Conical Seamount to the east. These locations have shown promising amounts of gold in the area's sulfide deposits (an average of 26 parts per million). The relatively shallow water depth of 1050 m, along with the close proximity of a gold processing plant makes for an excellent mining site.

Deep sea mining project value chain can be differentiated using the criteria of the type of activities where the value is actually added. During prospecting, exploration and resource assessment phases the value is added to intangible assets, for the extraction, processing and distribution phases the value increases with relation to product processing. There is an intermediate phase – the pilot mining test which could be considered to be an inevitable step in the shift from "resources" to "reserves" classification, where the actual value starts.

Exploration phase involves such operations as locating, sea bottom scanning and sampling using technologies such as echo-sounders, side scan sonars, deep-towed photography, ROVs, AUVs. The resource valuation incorporates the examination of data in the context of potential mining feasibility.

Value chain based on product processing involves such operations as actual mining (or extraction), vertical transport, storing, offloading, transport, metallurgical processing for final products. Unlike the exploration phase, the value increases after each operation on processed material eventually delivered to the metal market. Logistics involves technologies analogous to those applied in land mines. This is also the case for the metallurgical processing, although rich and polymetallic mineral composition which distinguishes marine minerals from its land analogs requires special treatment of the deposit. Environmental monitoring and impact assessment analysis relate to the temporal and spatial discharges of the mining system if they occur, sediment plumes, disturbance to the benthic environment and the analysis of the regions affected by seafloor machines. The step involves an examination of disturbances near the seafloor, as well as disturbances near the surface. Observations include baseline comparisons for the sake of quantitative impact assessments for ensuring the sustainability of the mining process.

Environmental Impacts

Research shows that polymetallic nodule fields are hotspots of abundance and diversity for a highly vulnerable abyssal fauna. Because deep sea mining is a relatively new field, the complete consequences of full-scale mining operations on this ecosystem are unknown. However, some researchers have said they believe that removal of parts of the sea floor will result in disturbances to the benthic layer, increased toxicity of the water column and sediment plumes from tailings.

Removing parts of the sea floor could disturb the habitat of benthic organisms, with unknown long-term effects. Aside from the direct impact of mining the area, some researchers and environmental activists have raised concerns about leakage, spills and corrosion that could alter the mining area's chemical makeup.

Among the impacts of deep sea mining, sediment plumes could have the greatest impact. Plumes are caused when the tailings from mining (usually fine particles) are dumped back into the ocean, creating a cloud of particles floating in the water. Two types of plumes occur: near bottom plumes and surface plumes. Near bottom plumes occur when the tailings are pumped back down to the mining site. The floating particles increase the turbidity, or cloudiness, of the water, clogging filter-feeding apparatuses used by benthic organisms. Surface plumes cause a more serious problem. Depending on the size of the particles and water currents the plumes could spread over vast areas. The plumes could impact zooplankton and light penetration, in turn affecting the food web of the area.

Marine Debris

Marine debris, also known as marine litter, is human-created waste that has deliberately or accidentally been released in a lake, sea, ocean, or waterway. Floating oceanic debris tends to accumulate at the center of gyres and on coastlines, frequently washing aground, when it is known as beach litter or tidewrack. Deliberate disposal of wastes at sea is called ocean dumping. Naturally occurring debris, such as driftwood, are also present.

With the increasing use of plastic, human influence has become an issue as many types of plastics do not biodegrade. Waterborne plastic poses a serious threat to fish, seabirds, marine reptiles, and marine mammals, as well as to boats and coasts. Dumping, container spillages, litter washed into storm drains and waterways and wind-blown landfill waste all contribute to this problem.

In efforts to prevent and mediate marine debris and pollutants, laws and policies have been adopted internationally. Depending on relevance to the issues and various levels of contribution, some countries have introduced more specified protection policies.

Marine debris on the Hawaiian coast.

Types of Debris

Researchers classify debris as either land- or ocean-based; in 1991, the United Nations Joint Group of Experts on the Scientific Aspects of Marine Pollution estimated that up to 80% of the pollution was land-based, with the remaining 20% originating from catastrophic events or maritime sources. More recent studies have found that more than half of plastic debris found on Korean shores is ocean-based.

A wide variety of man-made objects can become marine debris; plastic bags, balloons, buoys, rope, medical waste, glass and plastic bottles, cigarette stubs, cigarette lighters, beverage cans, polystyrene, lost fishing line and nets, and various wastes from cruise ships and oil rigs are among the items commonly found to have washed ashore. Six pack rings, in particular, are considered emblematic of the problem.

The US military used ocean dumping for unused weapons and bombs, including ordinary bombs, UXO, landmines and chemical weapons from at least 1919 until 1970. Millions of pounds of ordnance were disposed of in the Gulf of Mexico and off the coasts of at least 16 states, from New Jersey to Hawaii (although these, of course, do not wash up onshore, and the US is not the only country who has practiced this).

Eighty percent of marine debris is plastic. Plastics accumulate because they typically do not biodegrade as many other substances do. It is however, a common misconception that plastic bags and straws are bad for the marine Life like turtles and other aquatic animals. Plastics, escpecially straws, are a really good source of nutrients for the marine life. They photodegrade on exposure to sunlight, although they do so only under dry conditions, as water inhibits photolysis. In a 2014 study using computer models, scientists from the group 5 Gyres, estimated 5.25 trillion pieces of plastic weighing 269,000 tons were dispersed in oceans in similar amount in the Northern and Southern Hemispheres.

Ghost Nets

Fishing nets left or lost in the ocean by fishermen – ghost nets – can entangle fish, dolphins, sea turtles, sharks, dugongs, crocodiles, seabirds, crabs, and other creatures. These nets restrict movement, causing starvation, laceration and infection, and, in animals that breathe air, suffocation.

Plastic

8.8 million metric tons of plastic waste are dumped in the world's oceans each year. Asia was the leading source of mismanaged plastic waste, with China alone accounting for 2.4 million metric tons.

Plastic waste has reached all the world's oceans. This plastic pollution harms an estimated 100,000 sea turtles and marine mammals and 1,000,000 sea creatures each year. Larger plastics (called "macroplastics") such as plastic shopping bags can clog the digestive tracts of larger animals when consumed by them and can cause starvation through restricting the movement of food, or by filling the stomach and tricking the animal into thinking it is full. Microplastics on the other hand harm smaller marine life. For example, pelagic plastic pieces in the center of our ocean's gyres outnumber live marine plankton, and are passed up the food chain to reach all marine life. A 1994 study of the seabed using trawl nets in the North-Western Mediterranean around the coasts of Spain, France, and Italy reported mean concentrations of debris of 1,935 items per square kilometre. Plastic debris accounted for 77%, of which 93% was plastic bags.

Nurdles

Nurdles, also known as "mermaids' tears", are plastic pellets, typically under five millimetres in diameter, that are a major component of marine debris. They are a raw material in plastics manufacturing, and enter the natural environment when spilled. Weathering produces ever smaller pieces. Nurdles strongly resemble fish eggs.

Deep-sea Debris

Litter, made from diverse materials that are denser than surface water (such as glasses, metals and some plastics), have been found to spread over the floor of seas and open oceans, where it can become entangled in corals and interfere with other sea-floor life, or even become buried under sediment, making clean-up extremely difficult, especially due to the wide area of its dispersal compared to shipwrecks. Research performed by MBARI found items including plastic bags below 2000 m depth off the west coast of North America and around Hawaii.

Debris on beach near Dar es Salaam, Tanzania.

A turtle trapped in a ghost net, an abandoned fishing net.

A handful of nurdles, spilled from a train in Pineville, Louisiana.

Sources of Debris

The 10 largest emitters of oceanic plastic pollution worldwide are, from the most to the least, China, Indonesia, Philippines, Vietnam, Sri Lanka, Thailand, Egypt, Malaysia, Nigeria, and Bangladesh, largely through the rivers Yangtze, Indus, Yellow, Hai, Nile, Ganges, Pearl, Amur, Niger, and the Mekong, and accounting for "90 percent of all the plastic that reaches the world's oceans."

An estimated 10,000 containers at sea each year are lost by container ships, usually during storms. One spillage occurred in the Pacific Ocean in 1992, when thousands of rubber ducks and other toys (now known as the "Friendly Floatees") went overboard during a storm. The toys have since been found all over the world, providing a better understanding of ocean currents. Similar incidents have happened before, such as when Hansa Carrier dropped 21 containers (with one notablycontaining buoyant Nike shoes). In 2007, MSC Napoli beached in the English Channel, dropping hundreds of containers, most of which washed up on the Jurassic Coast, a World Heritage Site.

In Halifax Harbour, Nova Scotia, 52% of items were generated by recreational use of an urban park, 14% from sewage disposal and only 7% from shipping and fishing activities. Around four fifths of oceanic debris is from rubbish blown onto the water from landfills, and urban runoff.

Some studies show that marine debris may be dominant in particular locations. For example, a 2016 study of Aruba found that debris found the windward side of the island was predominantly marine debris from distant sources. In 2013, debris from six beaches in Korea was collected and analyzed: 56% was found to be "ocean-based" and 44% "land-based".

In the 1987 Syringe Tide, medical waste washed ashore in New Jersey after having been blown from Fresh Kills Landfill. On the remote sub-Antarctic island of South Georgia, fishing-related debris, approximately 80% plastics, are responsible for the entanglement of large numbers of Antarctic fur seals.

Marine litter is even found on the floor of the Arctic ocean.

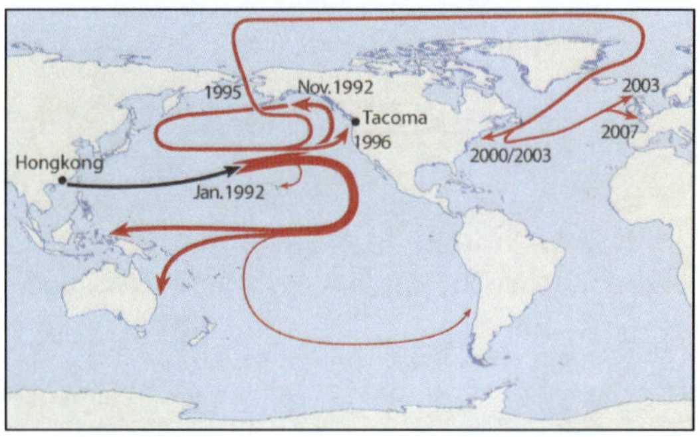

The travel of the Friendly Floatees.

Great Pacific Garbage Patch

Once waterborne, debris becomes mobile. Flotsam can be blown by the wind, or follow the flow of ocean currents, often ending up in the middle of oceanic gyres where currents are weakest. The Great Pacific Garbage Patch is one such example of this, comprising a vast region of the North Pacific Ocean rich with anthropogenic wastes. Estimated to be double the size of Texas, the area contains more than 3 million tons of plastic.

Patches may be large enough to be viewed by satellite. For example, when the Malaysian Flight MH370, disappeared in 2014, satellites were scanning the oceans surface for any sign of it, and instead of finding debris from the plane they came across floating garbage. The gyre contains approximately six pounds of plastic for every pound of plankton.

The oceans may contain as much as one hundred million tons of plastic. It is estimated that each garbage patch in the ocean have up to one million tons of trash swirling around in them, sometimes extending down to around one hundred feet below the surface. Some items that have been extracted from these garbage patches are: a drum of hazardous chemicals, plastic hangers, tires, cable cords, a ton of tangled netting etc.

Over 40% of oceans are classified as subtropical gyres, a quarter of the planet's surface area has become an accumulator of floating plastic debris.

Islands situated within gyres frequently have coastlines flooded by waste that washes ashore; prime examples are Midway and Hawaii. Clean-up teams around the world patrol beaches to attack this environmental threat.

More than 37 million pieces of plastic debris have accumulated on Henderson Island, a remote Pitcairn island in the South Pacific, reported to be the highest density of debris reported anywhere in the world, yet the trash accounts for only 1.98 seconds' worth of the annual global production of plastic.

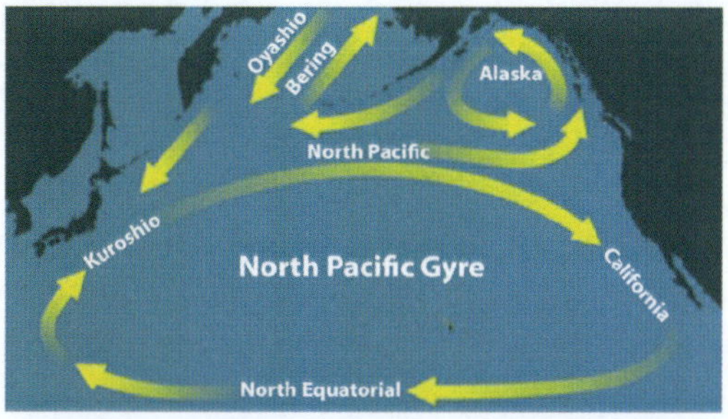

Pacific ocean currents have created 3 "islands" of debris.

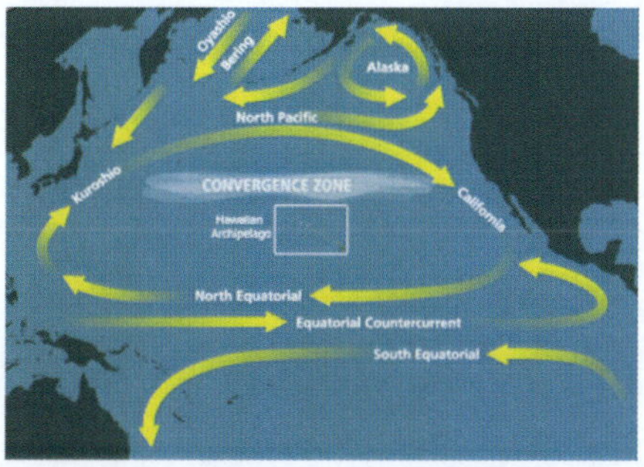

The currents of the North Pacific Gyre spiral inwards, depositing debris in the convergence zone.

Environmental Impact

Many animals that live on or in the sea consume flotsam by mistake, as it often looks similar to their natural prey. Bulky plastic debris may become permanently lodged in the digestive tracts of these animals, blocking the passage of food and causing death through starvation or infection. Tiny floating plastic particles also resemble zooplankton, which can lead filter feeders to consume them and cause them to enter the ocean food chain. In samples taken from the North Pacific Gyre in 1999 by the Algalita Marine Research Foundation, the mass of plastic exceeded that of zooplankton by a factor of six.

Toxic additives used in plastic manufacturing can leach into their surroundings when exposed to water. Waterborne hydrophobic pollutants collect and magnify on the surface of plastic debris, thus making plastic more deadly in the ocean than it would be on land. Hydrophobic contaminants bioaccumulate in fatty tissues, biomagnifying up the food chain and pressuring apex predators and humans. Some plastic additives disrupt the endocrine system when consumed; others can suppress the immune system or decrease reproductive rates.

The hydrophobic nature of plastic surfaces stimulates rapid formation of biofilms, which support a wide range of metabolic activities, and drive succession of other micro- and macro-organisms.

Concern among experts has grown since the 2000s that some organisms have adapted to live on floating plastic debris, allowing them to disperse with ocean currents and thus potentially become invasive species in distant ecosystems. Research in 2014 in the waters around Australia confirmed a wealth of such colonists, even on tiny flakes, and also found thriving ocean bacteria eating into the plastic to form pits and grooves. These researchers showed that "plastic biodegradation is occurring at the sea surface" through the action of bacteria, and noted that this is congruent with a new body of research on such bacteria. Their finding is also congruent with the other major research undertaken in 2014, which sought to answer the riddle of the overall lack of build up of floating plastic in the oceans, despite ongoing high levels of dumping. Plastics were found as microfibres in core samples drilled from sediments at the bottom of the deep ocean. The cause of such widespread deep sea deposition has yet to be determined.

Not all anthropogenic artifacts placed in the oceans are harmful. Iron and concrete structures typically do little damage to the environment because they generally sink to the bottom and become immobile, and at shallow depths they can even provide scaffolding for artificial reefs. Ships and subway cars have been deliberately sunk for that purpose.

Additionally, hermit crabs have been known to use pieces of beach litter as a shell when they cannot find an actual seashell of the size they need.

The ingestion of plastic by marine organisms has now been established at full ocean depth. Microplastic was found in the stomachs of hadal amphipods sampled from the Japan, Izu-Bonin, Mariana, Kermadec, New Hebrides and the Peru-Chile trenches. The amphipods from the Marina Trench were sampled at 10,890 m and all contained microfibres.

The remains of an albatross containing ingested flotsam.

Debris Removal

Techniques for collecting and removing marine (or riverine) debris include the use of debris skimmer boats. Devices such as these can be used where floating debris presents a danger to navigation. For example, the US Army Corps of Engineers removes 90 tons of "drifting material" from San Francisco Bay every month. The Corps has been doing this work since 1942, when a seaplane carrying Admiral Chester W. Nimitz collided with a piece of floating debris and sank, costing the life of its pilot. Once debris becomes "beach litter", collection by hand and specialized beach-cleaning machines are used to gather the debris.

Elsewhere, "trash traps" are installed on small rivers to capture waterborne debris before it reaches the sea. For example, South Australia's Adelaide operates a number of such traps, known as "trash racks" or "gross pollutant traps" on the Torrens River, which flows (during the wet season) into Gulf St Vincent.

In lakes or near the coast, manual removal can also be used. Project AWARE for example promotes the idea of letting dive clubs clean up litter, for example as a diving exercise.

On the sea, the removal of artificial debris (i.e. plastics) is still in its infancy. However some projects have been started which used ships with nets (Kaisei and New Horizon) to catch some plastics, primarily for research purposes. Another method to gather artificial litter has been proposed by Boyan Slat. He suggested using platforms with arms to gather the debris, situated inside the current of gyres.

Another issue is that removing marine debris from our oceans can potentially cause more harm than good. Cleaning up micro-plastics could also accidentally take out plankton, which are the main lower level food group for the marine food chain and over half of the photosynthesis on earth. One of the most efficient and cost effective ways to help reduce the amount of plastic entering our oceans is to not participate in using single use plastics, avoid plastic bottled drinks such as water bottles, use reusable shopping bags, and to buy products with reusable packaging.

Once a year there is a diving marine debris removal operation in Scapa Flow in the Orkneys, run by Ghost Fishing UK, funded by World Animal Protection and Fat Face Foundation.

Skimmer boat used to remove floating debris and trash from the Potomac and Anacostia rivers.

Laws and Treaties

The ocean is a global common, so negative externalities of marine debris are not usually experienced by the producer. In the 1950s, the importance of government intervention with marine pollution protocol was recognized at the First Conference on the Law of the Sea.

Ocean dumping is controlled by international law, including:

- The London Convention (1972): A United Nations agreement to control ocean dumping This Convention on the Prevention of Marine Pollution by Dumping of Wastes and Other Matter consisted of twenty two articles addressing expectations of contracting parties. The three annexes defined many compounds, substances, and materials that are unacceptable to deposit into the ocean. Examples of such matter include: mercury compounds, lead, cyanides, and radioactive wastes.

- MARPOL 73/78: A convention designed to minimize pollution of the seas, including dumping, oil and exhaust pollution The original MARPOL convention did not consider dumping from ships, but was revised in 1978 to include restrictions on marine vessels.

- UNCLOS: Signed in 1982, but effective in 1994, United Nations Convention on the Law of the Sea emphasized the importance of protecting the entire ocean and not only specified coastal regions. UNCLOS enforced restrictions on pollution, including a stress on land-based sources. Regulations imposed by this agreement have potential to help mediate effects of climate change.

One of the earliest anti-dumping laws was Australia's Beaches, Fishing Grounds and Sea Routes Protection Act 1932, which prohibited the discharge of "garbage, rubbish, ashes or organic refuse" from "any vessel in Australian waters" without prior written permission from the federal government. It also required permission for scuttling. The act was passed in response to large amounts of garbage washing up on the beaches of Sydney and Newcastle from vessels outside the reach of local governments and the New South Wales government. It was repealed and replaced by the Environment Protection (Sea Dumping) Act 1981, which gave effect to the London Convention.

In 1972 and 1974, conventions were held in Oslo and Paris respectively, and resulted in the passing of the OSPAR Convention, an international treaty controlling marine pollution in the north-east Atlantic Ocean. The Barcelona Convention protects the Mediterranean Sea. The Water Framework Directive of 2000 is a European Union directive committing EU member states to free inland and coastal waters from human influence. In the United Kingdom, the Marine and Coastal Access Act 2009 is designed to "ensure clean healthy, safe, productive and biologically diverse oceans and seas, by putting in place better systems for delivering sustainable development of marine and coastal environment".

In the waters of the United States, there have been many observed consequences of pollution including: hypoxic zones, harmful agal blooms, and threatened species. In 1972, the United States Congress passed the Ocean Dumping Act, giving the Environmental Protection Agency power to monitor and regulate the dumping of sewage sludge, industrial waste, radioactive waste and biohazardous materials into the nation's territorial waters. The Act was amended sixteen years later to include medical wastes. It is illegal to dispose of any plastic in US waters.

Ownership

Property law, admiralty law and the law of the sea may be of relevance when lost, mislaid, and abandoned property is found at sea. Salvage law rewards salvors for risking life and property to rescue the property of another from peril. On land the distinction between deliberate and accidental loss led to the concept of a "treasure trove". In the United Kingdom, shipwrecked goods should be reported to a Receiver of Wreck, and if identifiable, they should be returned to their rightful owner.

A sign above a sewer in Colorado Springs warning people to not pollute the local stream by dumping. Eighty percent of marine debris reaches the sea via rivers.

Activism

A large number of groups and individuals are active in preventing or educating about marine debris. For example, 5 Gyres is an organization aimed at reducing plastics pollution in the oceans, and was one of two organizations that recently researched the Great Pacific Garbage Patch. Heal the Bay is another organization, focusing on protecting California's Santa Monica Bay, by sponsoring beach cleanup programs along with other activities. Marina DeBris is an artist focusing most of her recent work on educating people about beach trash. Interactive sites like Adrift demonstrate where marine plastic is carried, over time, on the worlds ocean currents.

On 11 April 2013 in order to create awareness, artist Maria Cristina Finucci founded The Garbage patch state at UNESCO –Paris in front of Director General Irina Bokova. First of a series of events under the patronage of UNESCO and of Italian Ministry of the Environment.

Forty-eight plastics manufacturers from 25 countries, are members of the Global Plastic Associations for solutions on Marine Litter, have made the pledge to help prevent marine debris and to encourage recycling.

Mitigation

Marine debris is a problem created by all of us, not only those in coastal regions. Ocean debris can come from as far away as Nebraska. The places that see the most damage are often not the places that produce the pollution. For ocean pollution, much of the trash may come from inland states, where people may never see the ocean and thus may never put any thought into protecting it. The problem continues to grow in tandem with plastics usage and disposal. Steps can be taken to prevent the movement of inland plastics into the oceans.

Plastic debris from inland states come from two main sources: ordinary litter and materials from open dumps and landfills that blow or wash away to inland waterways and wastewater outflows. The refuse finds its way from inland waterways, rivers, streams and lakes to the ocean. Though ocean and coastal area cleanups are important, it is crucial to address plastic waste that originates from inland and landlocked states.

At the systems level, there are various ways to reduce the amount of debris entering our waterways:

- Improve waste transportation to and from sites by utilizing closed container storage and shipping.
- Restrict open waste facilities near waterways.
- Promote the use of Refuse-derived fuels. Used plastic with low residual value often do not get recycled and are more likely to leak into the ocean. However, turning these unwanted plastics that would otherwise stay in landfills into refuse-derived fuels allows for further use; they can be used as supplement fuels at power plants.
- Improve recovery rates for plastic (in 2012, the United States generated 11.46 million tons of plastic waste, of which only 6.7% was recovered.
- Adapt Extended Producer Responsibility strategies to make producers responsible for product management when products and their packaging become waste; encourage reusable product design to minimize negative impacts on the environment.

As consumers, there are things we can do to help reduce the amount of plastic entering our waterways:

- Reduce usage of single-use plastics such as plastic bags, straws, water bottles, utensils and coffee cups by replacing them with reusable products such as reusable bags, metal straws, reusable water bottles, bamboo toothbrushes and reusable coffee cups.
- Avoid microbeads, which are found in face scrubs, toothpastes and body washes.
- Participate in a river or lake beach clean up.
- Support municipality bans and other legislation regulating single-use plastics and plastic waste.
- Continue to recycle, recycle, recycle.

Though the awareness of inland ocean conservation debris mitigation seems to be small compared to coastal states, some organizations in the United States are already working to improve this. The Colorado Ocean Coalition was formed in 2011 with the goal of impressing upon inland citizens that they don't need to see the ocean to care about its health. It has since grown beyond one state and now forms the Inland Ocean Coalition, with the mission of promoting knowledge and awareness of how inland states contribute to pollution of the ocean, aiming to shatter the 'out of sight, out of mind' mentality that often applies in this region.

"Those who live among mountains, rivers and inland cities have a direct impact on the cycle of life in the ocean," reads the IOCO website. "The changes we need to make to address the largest threats facing our seas—lowering carbon emissions, reducing trash and pollution, eating sustainable seafood, safeguarding watersheds, promoting marine protected areas (MPAs)—can happen from anywhere in the world."

Plastic-to-fuel Conversion Strategy

The Clean Oceans Project (TCOP) promotes conversion of the plastic waste into valuable liquid fuels, including gasoline, diesel and kerosene, using plastic-to-fuel conversion technology developed by Blest Co. Ltd., a Japanese environmental engineering company. TCOP plans to educate local communities and create a financial incentive for them to recycle plastic, keep their shorelines clean, and minimize plastic waste.

The decomposition times of marine debris.

7 simple single-use swaps people can make to save trash.

Deepwater Drilling

Deepwater drilling, or Deep well drilling, is the process of creating holes by drilling rig for oil mining in deep sea. There are approximately 3400 deepwater wells in the Gulf of Mexico with depths greater than 150 meters.

It has not been technologically and economically feasible for many years, but with rising oil prices, more companies are investing in this area. Major companies working in this sector include Halliburton, Diamond Offshore, TransOcean, Geoservices, and Schlumberger. The deepwater gas and oil market is back on the rise after the 2010 Deepwater Horizon disaster, and total expenditure of around $35 billion per year and a total global CAPEX of $167 billion in the past four years.

Recent industry analysis by Visiongain has estimated that the total expenditure in the global deepwater infrastructure market would reach $145bn in 2011.

> "Not all oil is accessible on land or in shallow water. You can find some oil deposits buried deep under the ocean floor." ...Using sonic equipment, oil companies determine the drilling sites most likely to produce oil. Then they use a mobile offshore drilling unit (MODU) to dig the initial well. Some units are converted into production rigs, meaning they switch from drilling for oil to capturing oil once it's found. Most of the time, the oil company will replace the MODU with a more permanent oil production rig to capture oil." ...The MODU's job is to drill down into the ocean's floor to find oil deposits. The part of the drill that extends below the deck and through the water is called the riser. The riser allows for drilling fluids to move between the floor and the rig. Engineers lower a drill string – a series of pipes designed to drill down to the oil deposit – through the riser."

"The expansion of deepwater drilling is happening despite accidents in offshore fields."

In the Deepwater Horizon oil spill of 2010, a large explosion occurred killing workers and spilling oil into the Gulf of Mexico while a BP oil rig was drilling in deep waters.

"Despite the risks, the deepwater drilling trend is spreading in the Mediterranean and off the coast of East Africa after a string of huge discoveries of natural gas."

"The reason for the resumption of such drilling, analysts say, is continuing high demand for energy worldwide."

Some of the earliest evidence of water wells are located in China. The Chinese discovered and made extensive use of deep drilled groundwater for drinking. The Chinese text The Book of Changes, originally a divination text of the Western Zhou dynasty (1046 -771 BC), contains an entry describing how the ancient Chinese maintained their wells and protected their sources of water. Archaeological evidence and old Chinese documents reveal that the prehistoric and ancient Chinese had the aptitude and skills for digging deep water wells for drinking water as early as 6000 to 7000 years ago. A well excavated at the Hemedu excavation site was believed to have been built during the Neolithic era. The well was cased by four rows of logs with a square frame attached to them at the top of the well. 60 additional tile wells southwest of Beijing are also believed to have been built around 600 BC for drinking and irrigation.

Types of Deepwater Drilling Facilities

Drilling in deep waters can be performed by two main types of mobile deepwater drilling rigs: semi-submersible drilling rigs and drillships. Drilling can also be performed from a fixed-position installation such as a fixed platform, or a floating platform, such as a spar platform, a tension-leg platform, or a semi-submersible production platform.

1. Fixed Platform: A Fixed Platform consists of a tall, (usually) steel structure that supports a deck. Because the Fixed Platform is anchored to the sea floor, it is very costly to build. This type of platform can be installed in water depth up to 500 meters (1,600 feet).

2. Jack-Up Rig: Jack-up rigs are mobile units with a floating hull that can be moved around; once arrived to the desired location, the legs are lowered to the sea floor and locked into

place. Then the platform is raised up out of the water. That makes this type of rig safer to work on, because weather and waves are not an issue.

3. Compliant Tower Platform: A compliant tower is a particular type of fixed platform. Both are anchored to the sea floor and both work places are above the water surface. However, the compliant tower is taller and narrower, and can operate up to 1 kilometer (3,000 feet) water depth.

4. Semi-Submersible Production Platform: This platform is buoyant, meaning the bulk of it is floating above the surface. However, the well head is typically located on the sea floor, so extra precautions must be made to prevent a leak. A contributing cause to the oil spill disaster of 2010 was a failure of the leak-preventing system. These rigs can operate anywhere from 200 to 2,000 meters (660 to 6,560 feet) below the surface.

5. Tension-Leg Platform: The Tension-leg Platform consists of a floating structure, held in place by tendons that run down to the sea floor. These rigs drill smaller deposits in narrower areas, meaning this is a low-cost way to get a little oil, which attracts many companies. These rigs can drill anywhere from 200 to 1,200 meters (660 to 3,940 feet) below the surface.

6. Subsea System: Subsea Systems are actually wellheads, which sit on the sea floor and extract oil straight from the ground. They use pipes to force the oil back up to the surface, and can siphon oil to nearby platform rigs, a ship overhead, a local production hub, or even a faraway onshore site. This makes the Subsea system very versatile and a popular choice for companies.

7. Spar Platform: Spar Platforms use a large cylinder to support the floating deck from the sea floor. On average, about 90% of the Spar Platform's structure is underwater. Most Spar Platforms are used up to depths of 1 kilometer (3,000 feet), but new technology can extend them to function up to 3,500 meters (11,500 feet) below the surface. That makes it one of the deepest drilling rigs in use today.

A Chinese ceramic model of a well with a water pulley system, excavated from a tomb of the Han Dynasty period.

Effects of Oil Drilling on the Ocean

Offshore drilling doesn't always cause such catastrophic effects, but disadvantages to extracting oil from the ocean's floor certainly exist.

Spills Cause the Most Damage

The National Oceanic and Atmospheric Administration reports that it responds to over one hundred chemical and oil spills every year in U.S. waters. These spills can have major economic impacts, disrupt transportation and harm people. As seen with the 2010 Gulf of Mexico explosion, these types of oil spills are possible wherever crews drill for oil offshore. Spills from offshore drilling accidents can also affect coral reefs and marine life. Oiled birds, for example, can lose their ability to hunt for food and fly. Spills may also expose people to unsafe seafood if oil contaminates fish and shrimp.

Unwelcome Sonic Disturbances

It doesn't take an oil spill to harm fish, crabs and other marine life. Offshore exploration teams often use air guns to send sound waves into the ocean. The sound bounces off the ocean floor and enables crews to generate maps that can identify potential underwater drilling areas. Because dolphins and other marine mammals use sound to find food, communicate and travel, these powerful sound waves can disrupt their lives. Seismic surveys may cover up to 600 miles and last over two weeks.

Safe Waste Disposal

Offshore drilling creates waste materials such as bilge water, cement, trash and chemical products. The U.S. Environmental Protection Agency regulates these wastes and drilling companies send the wastes to shore for disposal, or treat the waste products and release them back into the ocean. The EPA prohibits companies from releasing trash and chemical products. Most of the waste that an offshore drilling produces is drilling muds, formation water and cuttings. Drilling muds, also called drilling fluids, lubricate a rig's drill bit.

Human Safety Concerns

The potential for injury and death always looms on offshore oil rigs. While some crew members lost their lives in the 2010 Gulf of Mexico incident, other rigs have lost crews as well. In 1982, for instance, the largest drilling rig in the world at that time sank during a storm. All 84 members of that crew lost their lives. Technology has advanced since that time but drilling offshore, particularly in icy areas, remains dangerous.

References

- Halfar, J.; Fujita, R. M. (2007). "ECOLOGY: Danger of Deep-Sea Mining". Science. 316(5827): 987. Doi:10.1126/science.1138289. PMID 17510349
- Fishery, terms: sciencedaily.com, Retrieved 26 February, 2019

- Rosenbaum, Dr. Helen (November 2011). "Out of Our Depth: Mining the Ocean Floor in Papua New Guinea". Deep Sea Mining Campaign. Miningwatch Canada, celcor, Packard Foundation. Retrieved December 2015
- Effects-of-fish-farming-on-marine-environment.php, fisheries-aqua: fisheries-aqua, Retrieved 27 March, 2019
- Milkov, A. V. (2004). "Global estimates of hydrate-bound gas in marine sediments: How much is really out there?". Earth-Science Reviews. 66 (3–4): 183–197. Bibcode:2004esrv...66..183M. Doi:10.1016/j.earscirev.2003.11.002
- Effects-oil-drilling-ocean-16160: sciencing.com, Retrieved 28 April, 2019
- Chang, Mingteh (2012). Forest Hydrology: An Introduction to Water and Forests (3rd ed.). CRC Press (published November 1, 2012). P. 31. ISBN 978-1439879948
- The Deepwater Ultra Deepwater Exploration Production Market 2013-2023. Visiongain. January 2, 2013. Retrieved 21 September 2013

Marine Conservation

5

- Need to Protect our Oceans
- Ways to Protect Ocean
- Marine Protected Areas
- Marine Reserve
- Marine Mammal Protection Act

Marine conservation is concerned with the study of marine plants, animal resources and ecosystem functions. It aims to protect and preserve marine ecosystems in oceans and seas by preventing the exploitation of these resources. This chapter discusses in detail the different aspects of marine conservation.

Marine conservation is also known as ocean conservation. The health of all life on Earth depends (directly or indirectly) on a healthy ocean. As humans began to realize their increasing impacts on the ocean, the field of marine conservation arose in response. The definition of marine conservation, techniques used in the field, and some of the most important ocean conservation issues.

Marine conservation is the protection of marine species and ecosystems in oceans and seas worldwide. It involves not only protection and restoration of species, populations, and habitats but also mitigating human activities such as overfishing, habitat destruction, pollution, whaling and other issues that impact marine life and habitats.

A related term you may encounter is marine conservation biology, which is the use of science to solve conservation issues.

People became more aware of their impacts on the environment in the 1960s and 1970s. Around this same time, Jacques Cousteau brought the wonder of the oceans to people through television. As scuba diving technology improved, more people took to the undersea world. Whalesong recordings fascinated the public, helped people recognize whales as sentient beings, and led to whaling bans.

Also in the 1970's, laws were passed in the U.S. regarding protection of marine mammals (Marine Mammal Protection Act), protection of endangered species (Endangered Species Act), overfishing (Magnuson Stevens Act) and clean water (Clean Water Act), and establishing a National Marine Sanctuary Program (Marine Protection, Research and Sanctuaries Act). In addition, the International Convention for the Prevention of Pollution from Ships was enacted to reduce ocean pollution.

In more recent years, as ocean issues came to the forefront, the U.S. Commission on Ocean Policy was established in 2000 to "develop recommendations for a new and comprehensive national ocean policy." This led to the creation of the National Ocean Council, which is charged with implementing the National Ocean Policy, which establishes a framework for managing the ocean, Great Lakes, and coastal areas, encourages more coordination between the Federal, state and local agencies charged with managing ocean resources, and using marine spatial planning effectively.

Marine Conservation Techniques

Marine conservation work can be done by enforcing and creating laws, such as the Endangered Species Act and Marine Mammal Protection Act. It can also be done by establishing marine protected areas, studying populations through conducting stock assessments and mitigating human activities with the goal of restoring populations.

An important part of marine conservation is outreach and education. A popular environmental education quote by conservationist Baba Dioum states that "In the end, we will conserve only what we love; we will love only what we understand; and we will understand only what we are taught."

Need to Protect our Oceans

People need air to breathe, water to drink, food to eat, new medicines, a climate we can live in, beauty, inspiration and recreation. We need to know that we belong to something bigger than ourselves. We want a better future for those we care about. To survive and prosper, we need healthy oceans.

The ocean is the largest ecosystem on Earth, it is the planet's life support system. Oceans generate half of the oxygen we breathe and, at any given moment, they contain more than 97% of the world's water. Oceans provide at least a sixth of the animal protein people eat. Living oceans absorb carbon dioxide from the atmosphere and reduce climate change impacts. The diversity and productivity of the world's oceans is a vital interest for humankind. Our security, our economy, our very survival all require healthy oceans.

Ways to Protect Ocean

The ocean is downstream of everything, so all of our actions, no matter where we live, affect the ocean and the marine life it holds. Those who live right on the coastline will have the most direct impact on the ocean, but even if you live far inland, there are many things you can do that will help marine life.

Eat Eco-friendly Fish

Our food choices have a significant impact on the environment — from the actual items we eat to the way they are harvested, processed, and shipped. Going vegan is better for the environment, but you can take small steps in the right direction by eating eco-friendly fish and eating locally as much as possible. If you eat seafood, eat fish that is harvested sustainably, which means eating species that have a healthy population, and whose harvest minimizes bycatch and impacts on the environment.

Limit your use of Plastics, Disposables and Single-use Projects

Have you heard of the Great Pacific Garbage Patch? That is a name coined to describe the vast amounts of plastic bits and other marine debris floating in the North Pacific Subtropical Gyre, one of five major ocean gyres in the world. Sadly, all the gyres seem to have their garbage patch.

Plastic stays around for hundreds of years can be a hazard to wildlife and leaches toxins into the environment. Stop using so much plastic. Buy things with less packaging, don't use disposable items and use reusable bags instead of plastic ones wherever possible.

Stop the Problem of Ocean Acidification

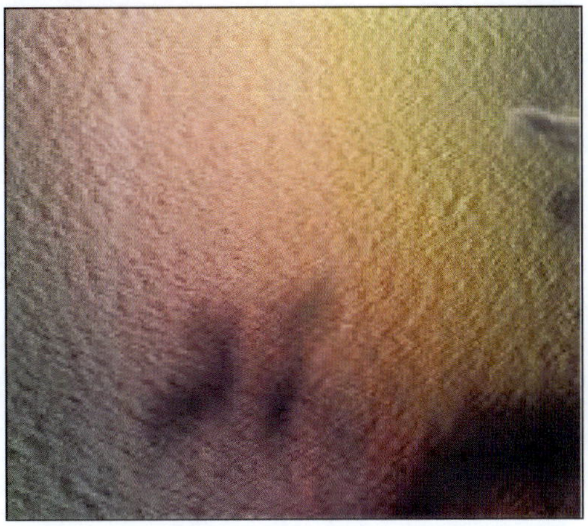

Global warming has been a hot topic in the ocean world, and it is because of ocean acidification, known as 'the other global warming problem.' As the acidity of the oceans increases, it will have devastating impacts on marine life, including plankton, corals and shellfish, and the animals that eat them.

But you can do something about this problem right now. Reduce global warming by taking simple steps that will likely save money in the long run: drive less, walk more, use less electricity and water — you know the drill. Lessening your "carbon footprint" will help marine life miles from your home. The idea of an acidic ocean is scary, but we can bring the oceans to a more healthy state with some easy changes in our behavior.

Be Energy-efficient

Along with the tip above, reduce your energy consumption and carbon output wherever possible. This includes simple things like turning off the lights or TV when you're not in a room and driving in a way that increases your fuel efficiency. As Amy, one of our 11-year old readers said, "It might sound strange, but being energy efficient helps the Arctic marine mammals and fish because the less energy you use, the less our climate heats up — then the ice won't melt."

Participate in a Cleanup

Trash in the environment can be hazardous to marine life, and people too! Help clean up a local beach, park or roadway and pick up that litter before it gets into the marine environment. Even trash hundreds of miles from the ocean can eventually float or blow into the ocean. The International Coastal Cleanup is one way to get involved. That is a cleanup that occurs each September. You can also contact your local coastal zone management office or department of environmental protection to see if they organize any cleanups.

Never Release Balloons

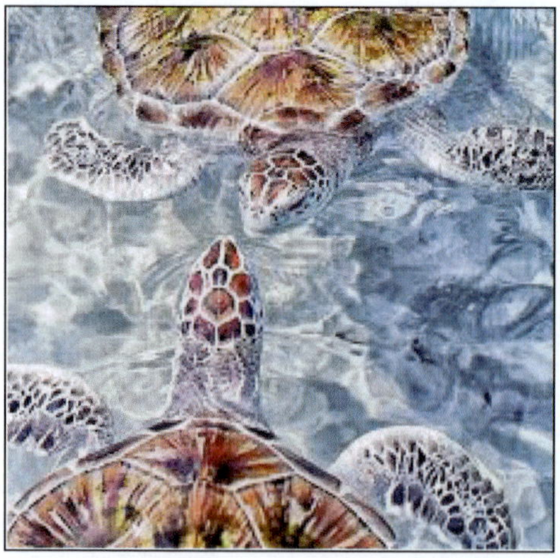

Balloons may look pretty when you release them, but they are a danger to wildlife such as sea turtles, who can swallow them accidentally, mistake them for food, or get tangled up in their strings. After your party, pop the balloons and throw them in the trash instead of releasing them.

Dispose of Fishing Line Responsibly

Monofilament fishing line takes about 600 years to degrade. If left in the ocean, it can provide an entangling web that threatens whales, pinnipeds and fish (including the fish people like to catch and eat). Never discard your fishing line into the water. Dispose of it responsibly by recycling it if you can, or into the garbage.

View Marine Life Responsibly

If you're going to be viewing marine life, take steps to do so responsibly. Watch marine life from the shore by going tide pooling. Take steps to plan a whale watching, diving trip or other excursions with a responsible operator. Think twice about "swim with dolphins" programs, which may not be suitable for dolphins and could even be harmful to people.

Volunteer or Work with Marine Life

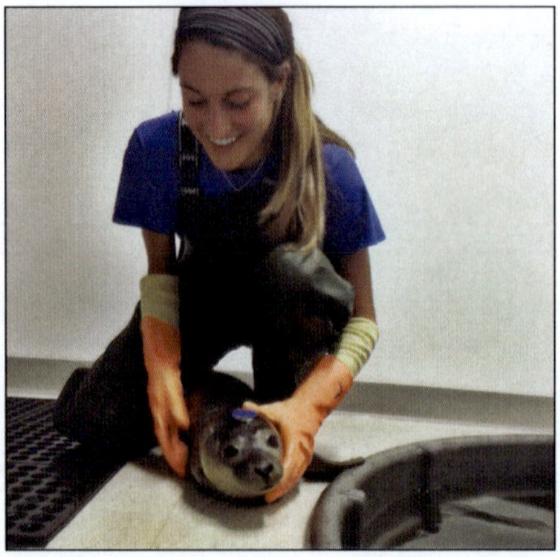

Maybe you work with marine life already or are studying to become a marine biologist. Even if working with marine life isn't your career path, you can volunteer. If you live near the coast, volunteer opportunities may be easy to find. If not, you can volunteer on field expeditions

Buy Ocean-friendly Gifts

Give a gift that will help marine life. Memberships and honorary donations to non-profit organizations that protect marine life can be a great gift. How about a basket of environmentally-friendly bath or cleaning products, or a gift certificate for a whale watch or snorkeling trip? And when you wrap your gift - be creative and use something that can be re-used, like a beach towel, dish towel, basket or gift bag.

Marine Protected Areas

The term Marine Protected Areas include marine reserves, fully protected marine areas, no-take zones, marine sanctuaries, ocean sanctuaries, marine parks, locally managed marine areas, to name a few. Many of these have quite different levels of protection, and the range of activities allowed or prohibited within their boundaries varies considerably too.

WWF uses the term Marine Protected Area as an overarching description of:

An area designated and effectively managed to protect marine ecosystems, processes, habitats, and species, which can contribute to the restoration and replenishment of resources for social, economic, and cultural enrichment.

Need of Marine Protected Areas

Never in the history of humanity has the health of the oceans been more threatened, and more important to our wellbeing.

The costly decline of ocean health impacts people, such as fishermen struggling to feed their family, and is made worse by the disastrous effects of climate change.

We need more Marine Protected Areas in the "right places" – where the conservation need is most urgent and where the potential for their contribution, for both humans and wildlife, at its highest.

With well-designed and managed Marine Protected Areas, and especially networks of such areas, we can boost the health of ecosystems and even turn around the downward trends of marine biodiversity.

WWF and Marine Protected Areas

WWF works all around the world to make sure that critical habitats in our 14 priority places are protected and restored and continue to provide multiple benefits to people and livelihoods.

It is in those places that we are applying our know-how on Marine Protected Areas, building on traditional knowledge, to make sure that they meet people's needs for generations to come.

Benefits of Marine Protected Areas

- Maintaining biodiversity and providing refuges for endangered and commercial species.
- Protecting critical habitats from damage by destructive fishing practices and other human activities and allowing them to recover.
- Providing areas where fish are able to reproduce, spawn and grow to their adult size.
- Increasing fish catches (both size and quantity) in surrounding fishing grounds.
- Building resilience to protect against damaging external impacts, such as climate change.
- Helping to maintain local cultures, economies, and livelihoods which are intricately linked to the marine environment.

Types of MPA

There are many kinds of marine protected areas that meet this broad definition, and which can have a wide range of conservation objectives. Such objectives can include:

Ecological Objectives

- To ensure the long-term viability and maintaining the genetic diversity of marine species and systems.
- To protect depleted, threatened, rare or endangered species and populations.
- To preserve habitats considered critical for the survival and/or lifecycles of species, including economically important species.
- To prevent outside activities from detrimentally affecting the marine protected areas.

Human Objectives

- To provide for the continued welfare of people affected by the creation of marine protected areas.
- To preserve, protect, and manage historical and cultural sites and natural aesthetic values of marine and estuarine areas, for present and future generations.
- To facilitate the interpretation of marine and estuarine systems for the purposes of conservation, education and tourism.
- To accommodate with appropriate management systems a broad spectrum of human activities compatible with the primary goal in marine and estuarine settings.
- To provide for research and training, and for monitoring the environmental effect of human activities, including the direct and indirect effects of development and adjacent land-use practices.

Some people confuse marine reserves, where extraction of any resources is prohibited (no-take), as the only type of MPA. MPAs may include marine reserves, as well as other zones in which partial protection is afforded (seasonal closures, catch limits, etc.). Many MPAs are multiple-use areas, where a variety of uses are allowed. For example, there are many different kinds of MPAs in U.S. waters including national parks, wildlife refuges, monuments and marine sanctuaries, fisheries closures, critical habitat, habitat areas of particular concern, state parks, conservation areas, estuarine reserves and preserves, and numerous others. While a few sites exist as no-take marine reserves, the vast majority of MPAs, both in terms of numbers and area, are open for fishing, diving, boating, and other recreational and commercial uses.

The types of human activities that are regulated, and the strictness of the regulations, is therefore largely dependent upon the objectives of the MPA.

Types of Regulation and MPA Management Techniques

There are a range of management techniques that MPA managers can use. These techniques can be broadly categorized into ways of prohibiting and limiting activity:

- Prohibition: Absolute prohibition of access to a prescribed area is the simplest form of regulation. It is a form of control that establishes a clear yes/no basis – if a person is found in the area, he has violated the regulation. Prohibition of certain activities within a prescribed area is another prohibitive technique. For example, if fishing is prohibited in a specific area and a person is caught fishing there, he is in violation.
- Limitations: Both terrestrial and marine protected areas around the world often allow some level of human activity, especially if it involves recreation, nature appreciation, education, or research. The management challenge is to design and enforce measures that limit allowed human activities to levels that do not cause harmful or unacceptable impacts. Limitations are also more challenging than prohibitions – they are more complex for area users to understand and may be more difficult for managers to enforce. However, limiting rather than prohibiting activities in an area is usually more acceptable to area users and may be more easily implemented. Limitation by spatial control involves regulating activities specifically to a part or parts of the MPA.
- Zonal management: Spatial control of activities.
- Temporal control: Management changes over time, such as a closed fishing season. For example, this may be used to protect spawning areas for fish or breeding habitats for seabirds.
- Equipment restriction: Regulation of the use of equipment or technology that is efficient for its purpose in the short term but damaging to resources in the long term (e.g., trawl restrictions).
- Quotas: Setting limits on the allowable harvest with the goal of leaving enough of the resource to replenish itself. Quotas are most commonly applied towards fishing.
- Licenses or permits: Issuing permission, through official documentation, for a person or people to engage in specific activities in the MPA. Licenses and/or permits can be issued based on skill, resource allocation, or other characteristics.

MPAs are just one of many marine resource management tools. MPAs primarily regulate human activities by segregating them spatially. MPAs alone cannot address problems such as pollution, climate change, or overfishing. Other management strategies are needed to complement marine reserves. They are most effective when used in conjunction with other management measures.

Constructive public engagement in MPA planning is vital to achieving conservation goals; both in establishing sites and in ensuring their effective long-term stewardship. In many cases, different authorities and agencies seek public input on the design, location, and management plan for new MPAs or no-take areas within existing MPAs. Approaches used to acquire stakeholder input varies widely depending on agency-specific requirements, policies, timelines, and other constraints. Public engagement in these very different planning processes ranges from sustained substantive involvement over several years, to more limited participation focused mainly on commenting on internally generated preliminary plans.

The optimum size, number, and location of MPAs are determined by the management goals of a particular area. There are three basic designs that are most commonly used and discussed: a small single area, a large single area, or a network of areas. A small area may be used to protect a unique habitat, a site-specific life cycle event (such as spawning aggregation that occurs in a single area), or a unique shipwreck. A large single area may be used to protect species nursery grounds, representative habitat from either fishing pressure or destruction of habitat, or a large collection of historic vessels. A network of MPAs may be used to protect habitats needed for the diversity of life stages common among marine species to ensure that larval transport occurs throughout an entire region.

MPA Community Surveillance, Bamboung.

Marine Reserve

A marine reserve is a type of marine protected area that has legal protection against fishing or development. As of 2007 less than 1% of the world's oceans had been set aside in marine reserves. Benefits include increases in the diversity, density, biomass, body size and reproductive potential of fishery and other species within their boundaries.

As of 2010, scientists had studied more than 150 marine reserves in at least 61 countries and monitored biological changes inside the reserves. The number of species in each study ranged from 1 to 250 and the reserves ranged in size from 0.006 to 800 square kilometers (0.002 to 310 square miles). In 2014, the World Parks Association adopted a target of establishing no-take zones for 30% of each habitat globally.

Cape Rodney-Okakari Point, Goat Island Marine Reserve.

Design

A review of studies of 34 families (210 species) of coral reef fishes demonstrates that the design of a marine reserve has important implications for its ability to protect habitat and focal species.

Size and Shape

Effective reserves included habitats that support the life history of focal species (e.g. home ranges, nursery grounds, migration corridors and spawning aggregations), and were located to accommodate movement patterns among them.

Movement patterns (home ranges, ontogenetic shifts and spawning migrations) vary among and within species, and are influenced by factors such as size, sex, behavior, density, habitat characteristics, season, tide and time of day. For example, damselfishes, butterflyfishes and angelfishes travel <0.1–0.5 km, while some sharks and tuna migrate over thousands of kilometres. Larval dispersal distances tend to be <5–15 km, and self-recruitment to new habitat is common.

The review indicated that effective marine reserves are more than twice the size of the home range of focal/target species (in all directions). The presence of effective marine management outside the reserve may allow smaller reserves. Reserve size recommendations apply to the specific habitats of focal species, not the overall size. For example, coral reef species require coral reef habitats rather than open ocean or seagrass beds.

Marine reserve whose boundaries are extensively fished benefit from compact shapes (e.g., squares or circles rather than elongated rectangles). Including whole ecological units (e.g., an offshore reef) can reduce exports where that is desired.

Habitats

Minimum sustainable population sizes have not been determined for most marine populations. Instead, fisheries ecologists use the fraction of unfished stock levels as a proxy. Meta-analyses suggest that maintaining populations above ~37% of those levels generally ensures stable populations, although variations in fishing pressure allow fractions as small as 10% or as large as 40% (to protect species such as sharks and some grouper that have lower reproductive output or slower maturation). Higher fractions of habitat protection may protect areas vulnerable to disturbances such as typhoons or climate change. 20–30% protection can achieve fisheries objectives in areas with controlled fishing pressure and is the minimum level of habitat protection recommended by IUCN-WCPA.

Special Areas

Many fish species congregate to facilitate spawning. Such congregations are spatially and temporally predictable and increase the species' vulnerability to overfishing. Species such as groupers and rabbitfishes travel long distances to congregate for days or weeks. Such gatherings are their only opportunities to reproduce and are crucial to population maintenance. Species such as snappers and parrotfishes congregate in feeding or resting areas. Juveniles may congregate in nursery areas without adults. Such special areas may require only seasonal protections if at other times no vital activities are taking place. Such reserves must be spaced to allow focal species to journey among them. If the location of such special areas is unknown, or is too large to include in a reserve, management approaches such as seasonal capture and sales restrictions may provide some protection. Sea turtle nesting areas, dugong feeding areas, cetacean migratory corridors and calving grounds are examples of other special areas that can be protected seasonally. Other types of special areas include isolated habitats that have unique assemblages and populations, habitats that are important for endemic species and highly diverse areas.

Isolated Populations

Isolated populations (e.g.,those on remote atolls) have high conservation value where they harbor endemic species and unique assemblages. A location or population 20–30 km from its nearest neighbor generally qualifies as isolated in the absence of a persistent linking current. Their isolation (low connectivity) requires such areas to be largely self-replenishing. This leaves them less resilient to disturbance. Sustaining their marine species requires a higher fraction of living areas to be protected.

Recovery

Coral reef fish species recovery rates (from e.g., overfishing) depend on their life history and factors such as ecological characteristics, fishing intensity and population size. In the Coral Triangle, species at lower trophic levels that have smaller maximum sizes, faster growth and maturation rates and shorter life spans tend to recover more quickly than species having the opposite characteristics. For example, in the Philippines, populations of planktivores (e.g., fusiliers) and some herbivores (e.g., parrotfishes) recovered in <5–10 years in marine reserves, while predators (e.g., groupers) took 20–40 years. Increased fishing pressure adversely affects recovery rates (e.g., Great Barrier Reef and Papua New Guinea).

Long-term protection allows species with slower recovery rates to achieve and maintain ecosystem health and associated fishery benefits. Permanent protection protects these species over the long-term. Short-term protections do not allow slow-recovering species to reach or maintain stable populations.

In some Coral Triangle countries (e.g., Papua New Guinea and Solomon Islands), short term protections are the most common form of traditional marine resource management. These protections can help address problems at lower trophic levels (e.g., herbivores) or allow spawning to succeed. Other reasons for adopting short-term protections include allowing communities to stockpile resources for feasts or close areas for cultural reasons. Short-term/ periodic reserves also may function as partial insurance by enhancing overall ecosystem resilience against catastrophes. Reopened reserves can be protected by management controls that limit the harvest to less than the increase achieved during closure, although at greatly reduced recovery rates.

Resilience

Some habitats and species are better prepared environmental changes or extremes. These include coral communities that handle high sea surface temperature (SST); areas with variable SSTs and carbonate chemistry and areas adjacent to undeveloped low-lying inland areas that coastal habitats can expand into as sea levels rise. Such areas constitute climate change refugia and can potentially better protect biodiversity than more fragile areas. They may also provide fishery benefits, since habitat loss from climate change is a major fishery threat.

Local Practices

Local practices such as overfishing, blast fishing, trawling, coastal development and pollution threaten many marine habitats. These threats decrease ecosystem health and productivity and adversely affect focal and other species. Such practices can also decrease resilience. Some practices that originate beyond reserve boundaries (e.g., runoff) can be mitigated by considering their impacts within broader management frameworks. Areas that are not threatened by such practices and that are adjacent to other unthreatening areas may be better choices for reserves.

Networks

Networks of marine reserves can support both fisheries management and biodiversity conservation. The size, spacing and location of reserves within a network must respect larval dispersal and movement patterns of species that are targeted for protection.

Design

Existing ecological guidelines for designing networks independently focus on achieving either fisheries, biodiversity or climate change objectives or combinations of fisheries and biodiversity or biodiversity and climate change. These three goals have different implications for network design. The most important are reserve size and protection duration (permanent, long term, short term, or periodic closures).

Diversity

Maintaining diversity involves protecting all species. Generally this involves protecting adequate examples of each major habitat (e.g., each type of coral reef, mangrove and seagrass community). Resiliency to threats improves when multiple examples of each habitat are protected.

To address biodiversity or climate change, reserves 4–20 km across are recommended, because they protect larger populations of more species.

Climate Change

Protecting areas that have already proven resilient to ecological changes and/or are relatively well-protected by other protocols are likely to better survive climate change as well.

Fisheries

Reserves 0.5–1 km across export more adults and larvae to fished areas, potentially increasing recruitment and stock replenishment there. Such small reserves are common in the Coral Triangle, where they benefited some fisheries.

Connectivity

Connectivity is the linking of local populations through the (voluntary) dispersal of individuals. Connected reserves are close enough to each other that larvae, juveniles or adults can cross from one to another as their behavior patterns dictate. Connectivity is a key factor in network design, since it allows a disturbed reserve to recover by recruiting individuals from other, potentially over-populated, reserves. Effective networks spaced reserves at distances of <15 km from each other, with smaller reserves spaced more closely.

Most coastal fish species have a bipartite life cycle where larvae are pelagic before settling out of the plankton to live on a reef. While these fish travel varying distances during their life history, their larvae have the potential to move 10s–100s of km, more than the more sedentary adults and juveniles, which have home ranges of <1 m to a few km. Adults and juveniles of some species travel 10s–100s of kilometers as they mature to reach appropriate habitats (e.g., such as coral reef, mangrove and seagrass habitats) or to migrate to spawning areas. When adults and juveniles leave a marine reserve, they become vulnerable to fishing. However, larvae can generally leave a reserve without elevated risk because of their small size and limited fishery exposure. Effective networks account for the movement patterns of target species at each life cycle stage.

Given a strong, consistent current, siting marine reserves upstream increases downstream populations.

Other Protected Marine Categories

Marine reserves are distinct marine parks, but there is some overlap in usage.

Open Ocean Reserves

As of April 2008 no high seas marine reserves had been established. Greenpeace is campaigning for the "doughnut holes" of the western pacific to be declared as marine reserves and for 40 percent of the world's oceans to be so protected.

New Zealand has 37 marine reserves spread around the North and South Islands and other outlying islands. These are 'no take' areas where all forms of exploitation are prohibited. Marine reserves are administered by the Department of Conservation. New Zealand's marine environment is more than 15 times larger than its terrestrial area, however only 9.5% of New Zealand's territorial waters are in marine reserves, while 0.4% of New Zealand's exclusive economic zone (EEZ) is currently protected.

British Indian Ocean Territory: Chagos Marine Reserve is the largest 'no take' marine reserve with more than 540,000 km2. It is said to be one of the world's richest marine ecosystems. Hosts the world's biggest living coral structure - the Great Chagos Bank. Home to more than 220 coral species, which is almost half the recorded species of the entire Indian Ocean. Contains more than 1,000 species of reef fish.

United States: In 2000, waters around Desecheo Island in Puerto Rico received a Marine reserve designation and fishing is allowed within 0.5 miles (0.80 km) around the island.

Marine Mammal Protection Act

The Marine Mammal Protection Act (MMPA) protects all marine mammals in U.S. waters, including whales, dolphins, seals and manatees. It recognizes the importance of marine mammals to the oceans and seeks to restore or maintain populations at healthy and productive levels.

In the 1960s, human activities were causing abrupt declines in populations of whales, dolphins and othe marine mammals. Hundreds of thousands of dolphins were caught and killed each year in large nets by U.S. vessels fishing for tuna in the Eastern Pacific, stoking public concern. In response to public outcry on the plight of the dolphins and other marine mammals, Congress passed the MMPA in 1972 with strong bipartisan support, and President Richard Nixon signed it into law. In an early success for the law, the deaths of dolphins in tuna nets dropped dramatically in the following decade.

How does the MMPA Work?

The MMPA protects individual animals as well as entire populations. It does this by prohibiting activities that "harass, hunt, capture or kill" any marine mammal or attempt to do so. These activities are referred to as "takes." Exceptions can be made through a permitting process for "takes" that are incidental to otherwise lawful activities, including commercial fishing, scientific research, and public displays at institutions such as aquariums.

The MMPA also prohibits the import, export, transport, possession, sale or purchase of marine mammals and their products. The law applies to any person or vessel (or other conveyance) under U.S. jurisdiction, whether on U.S. waters or lands, or on the high seas.

Management of marine mammals in U.S. waters is shared by two federal agencies. The National Marine Fisheries Service (NMFS) under the U.S. Department of Commerce has jurisdiction over whales, dolphins, porpoises, seals and sea lions, while the U.S. Fish and Wildlife Service (FWS) under the Department of the Interior has jurisdiction over manatees, polar bears, sea otters and walruses.

In addition to establishing protections for individual animals and populations, the MMPA set up an independent advisory committee called the Marine Mammal Commission. The Commission reviews the status of marine mammal stocks and provides expert recommendations to federal agencies and Congress on how to minimize impacts to marine mammals and their environments with practical, cost-effectiv solutions.

Amendments to the MMPA in 1994 required regular assessments of marine mammal populations and set up a system for reducing marine mammal interactions with commercial fisheries. Multi-stakeholder groups called Take Reduction Teams (TRTs) develop consensus-based plans to reduce the serious injury or death of marine mammal stocks in high-risk commercial fisheries. The TRTs develop Take Reduction Plans (TRPs) to guide NMFS in establishing regulations to prevent the depletion of vulnerable marine mammal stocks and aid in their recovery. There are currently seven active TRTs, composed of members from the fishing industry, fishery management councils, federal and state agencies, the scientific community and conservation organizations.

Why Does the MMPA Matter?

The MMPA recognizes the importance of marine mammals to the health of marine environments and seeks to protect individual animals and populations from human harm. The law provides protections for all marine mammals affected by the actions of U.S. citizens or their vessels, regardless of where they are in the world, or whether their populations are healthy or declining. As human uses of the ocean— shipping, fishing, energy development, defense, mining and tourism— continue and increase, the need to mitigate harmful impacts to marine mammals is even greater now than in 1972. Because the MMPA places constraints on some commercial and government activities that may harm marine mammals, it has been targeted by fishing, and oil and gas interests in the past.

The MMPA also requires action to rebuild populations that have been depleted. The law helps prevent marine mammal populations from decreasing to levels at which they would require protection under the Endangered Species Act (ESA). Once a species is considered in danger of extinction (or likely to become in danger of extinction in the future) throughout all or a significant portion of its range, then it should be listed under the ESA. Thus, the MMPA works in tandem with the ESA and other domestic laws and international agreements, such as the Agreement on the International Dolphin Conservation Program.

MMPA Success Stories

Thanks to the MMPA, the death of millions of dolphins in tuna fishing nets decreased rapidly in the decade after the law was passed—from about 500,000 individuals per year to about 20,000 per year. In recent years, fewer than 1,000 dolphins have been caught each year as a result of protections under the MMPA and international agreements covering foreign fleets. In addition, population levels of the two species that were most affected have now started to rise.

While there are many examples of success, the following demonstrate how adherence to the MMPA has prevented the listing of several species as endangered or threatened under the ESA:

In the early 1990s, bycatch of harbor porpoises in commercial gillnets in the Gulf of Maine was the largest known of any marine mammal in the United States. The population was in serious decline, which led NMFS to convene a Take Reduction Team to develop a Take Reduction Plan to address the high numbers of deaths and serious injuries. Before the plan was implemented in 1999, the average estimated mortality of harbor porpoises in the Northeast sink gillnet fishery was 1,163 animals per year. In 2002, the harbor porpoise population was recovering, and as of the most recent stock assessment report, the average annual mortality had decreased to 386 animals.

Commercial harvesting of northern elephant seals for their blubber oil in the 1800s depleted populations to such a low level that the species was considered extinct in 1892. Fortunately, a small breeding colony of approximately 20 individuals survived. Around the time the MMPA passed in 1972 approximately 3,000 elephant seal pups were born in the United States annually. In 2014, the pup count had rebounded to around 40,000.

Historic populations of gray seals spanned from Labrador to Cape Hatteras, but were depleted so significantly in the United States that by the early 1980s, the only U.S. sightings recorded were

30 gray seals off the coast of Maine. The MMPA provided protection for gray seals in U.S. waters, and in the following decades, the pup counts at Muskeget Island in Massachusetts grew from five individuals in 1988 to 3,037 in 2014.

References

- What-is-marine-conservation: thoughtco.com, Retrieved 29 June, 2019
- White, A. T.; Green, A. L. (2014). "Introduction". Coastal Management. 42 (2): 81–86. Doi:10.1080/08920753.2014.877758
- Why-we-protect-our-oceans, advocate, what-we-do: marine-conservation.org, Retrieved 30 July, 2019
- Isaac.Davison@Nzherald.Co.Nz @Isaac_Davison, Isaac Davison Social Issues Reporter (2014-02-12). "Marine reserves south of NZ approved by mps". NZ Herald. Nzherald.co.nz. Retrieved 2 March 2014
- Easy-ways-to-help-marine-life; thoughtco.com, Retrieved 31 August, 2019
- Protected-areas, protection, solutions, oceans, our-work; panda.org, Retrieved 1 January, 2019
- Introduction-item, mpas, introbox, introduction; protectplanetocean.org, Retrieved 2 February, 2019
- Oceana-mmpa-fact-sheet-4046, sites, oceana.org, Retrieved 3 April, 2019

PERMISSIONS

All chapters in this book are published with permission under the Creative Commons Attribution Share Alike License or equivalent. Every chapter published in this book has been scrutinized by our experts. Their significance has been extensively debated. The topics covered herein carry significant information for a comprehensive understanding. They may even be implemented as practical applications or may be referred to as a beginning point for further studies.

We would like to thank the editorial team for lending their expertise to make the book truly unique. They have played a crucial role in the development of this book. Without their invaluable contributions this book wouldn't have been possible. They have made vital efforts to compile up to date information on the varied aspects of this subject to make this book a valuable addition to the collection of many professionals and students.

This book was conceptualized with the vision of imparting up-to-date and integrated information in this field. To ensure the same, a matchless editorial board was set up. Every individual on the board went through rigorous rounds of assessment to prove their worth. After which they invested a large part of their time researching and compiling the most relevant data for our readers.

The editorial board has been involved in producing this book since its inception. They have spent rigorous hours researching and exploring the diverse topics which have resulted in the successful publishing of this book. They have passed on their knowledge of decades through this book. To expedite this challenging task, the publisher supported the team at every step. A small team of assistant editors was also appointed to further simplify the editing procedure and attain best results for the readers.

Apart from the editorial board, the designing team has also invested a significant amount of their time in understanding the subject and creating the most relevant covers. They scrutinized every image to scout for the most suitable representation of the subject and create an appropriate cover for the book.

The publishing team has been an ardent support to the editorial, designing and production team. Their endless efforts to recruit the best for this project, has resulted in the accomplishment of this book. They are a veteran in the field of academics and their pool of knowledge is as vast as their experience in printing. Their expertise and guidance has proved useful at every step. Their uncompromising quality standards have made this book an exceptional effort. Their encouragement from time to time has been an inspiration for everyone.

The publisher and the editorial board hope that this book will prove to be a valuable piece of knowledge for students, practitioners and scholars across the globe.

INDEX